A BETTER BOOK SERIES

BUILD YOUR OWN HOME

A GUIDE FOR SUBCONTRACTING THE EASY WAY

A SYSTEM TO SAVE TIME AND MONEY — ILLUSTRATED

BY J. JAMES HASENAU

The author, a second generation builder, has described the different operations in sub-contracting a house, recreation room or addition, disclosing the short cuts and the various ways to accomplish them; what to look out for in advantages and disadvantages of some systems used in construction. He describes where lines are drawn between the tradesmen, what they are responsible for and what is the next man's job. No attempt is made to tell how to pound nails or do a subs job, only what work he should do, from a builder's standpoint.

Covered in detail is the whole story of building a house from the builder's view of acquiring the land; plans; financing; insurance; paying the men, and inspections. It is simplified enough so that anyone should be able to follow it.

For the do-it-yourself builder, many manufacturers are producing parts of houses which are easy to use and go together with a minimum amount of skill. Plastic plumbing needs no torch or leaded joints; aluminum windows need only to be leveled up and nailed in place. The savings are tremendous.

Professional builders hire sub-contractors to do the work. Each sub is a business man, owner and mechanic with a crew of men, or sometimes he just works by himself. The organizing of a groups of men to build a house is outlined in this book. There are many possibilities of mistakes in building, and it is the hope of the author that these chapters will guide the amateur through the troubled spots, or at least let him know what the professional builder is up against and how he earns his money.

The individual can organize the building program himself, and save money in many ways avoiding costly errors in following chapter by chapter the outline of jobs for each sub-contractor in the book.

It pictures the system as the author has experienced it and how it works. Whether it works or not for the amateur depends upon many circumstances, but it is certainly better than going in cold without any advice or help.

INTRODUCTION

In the years of experience I have had in building, I have built many homes for people who tried it once and got too many gray hairs. I also finished many homes that people started and could not finish.

Today's contractor earns a fair profit in any type of building operation. If you are a good mechanic and fair organizer, you may be able to save money and do some of the work yourself, by acting as your own builder, or prime contractor. This book outlines the order in which the subcontractors hired by the builder should do their work, where one sub leaves off and another starts, and the gray area between one sub and another.

The ordering of building materials, paying procedures, obtaining financing, setting grades and most of the problems of the professional builder are covered in the outline.

No attempt is made to describe how to do any subcontractor's job. Instead, the book tells you what to watch for as each does his job, and how to instruct him. Different systems of construction and material use are presented, with advantages and disadvantages explained.

Examples of possible costly mistakes are mentioned so that home owners and builders can avoid them, and cut down the building risk factor as much as possible.

BUILD YOUR OWN HOME

A GUIDE FOR SUBCONTRACTING THE EASY WAY

A System to Save Time and Money

MAJOR Subcontractors and the order of what they do.

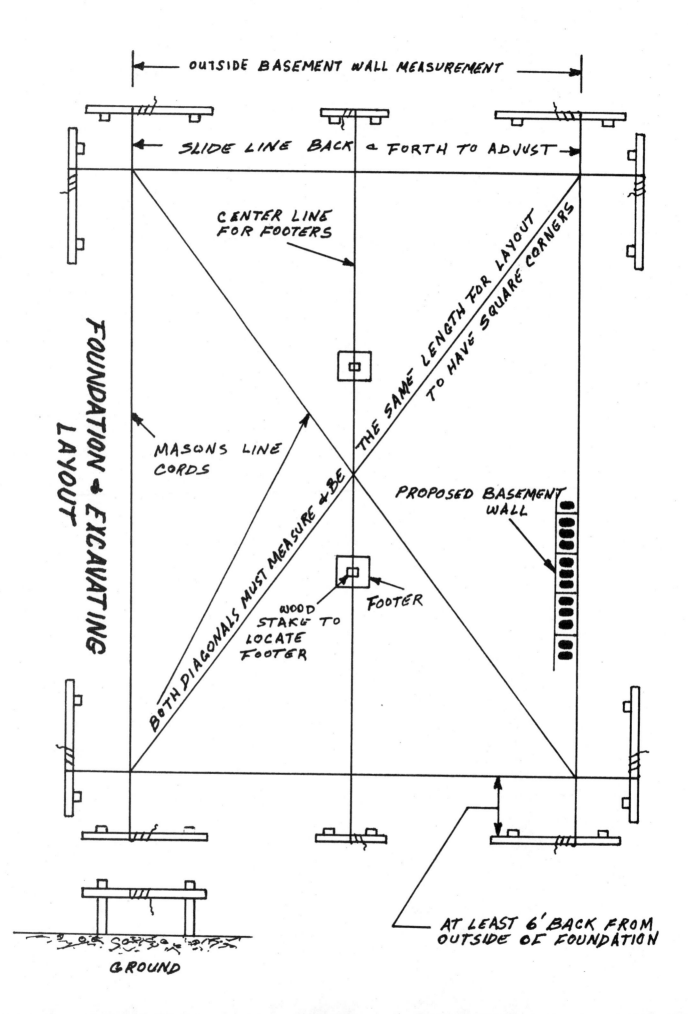

OUTSIDE BASEMENT WALL MEASUREMENT

SLIDE LINE BACK & FORTH TO ADJUST

CENTER LINE FOR FOOTERS

THE SAME LENGTH FOR LAYOUT TO HAVE SQUARE CORNERS

FOUNDATION & EXCAVATING LAYOUT

MASONS LINE CORDS

BOTH DIAGONALS MUST MEASURE & BE

PROPOSED BASEMENT WALL

WOOD STAKE TO LOCATE FOOTER

FOOTER

AT LEAST 6' BACK FROM OUTSIDE OF FOUNDATION

GROUND

1

PROFESSIONAL SUBCONTRACTOR

In building, the professional subcontractor enters where the operation is very exact and requires much training. The divisions of labor are quite distinct — brick layer, carpenter, cement man and many more, in fact some 40 trades in all are required.

Most books of this nature low-ball the professionals and convey that their jobs can be done by anyone with a little handyman ability. This attitude is frightening, for today's building trades constitute a mass of knowledge, manipulative ability, and years of on-the-job experience. In the trades, one hears the word 'fudge' often. This means making something look right when it is not perfect. Very few rooms are perfectly square, very few walls exactly plumb. A tile setter, for example, must know how to correct the eighth-inch variation between the top and bottom rows of tile when the enclosing walls are off that much, and they usually are.

The block layer must know how to give the carpenter a good base to put the house on, even though the blueprint does not show every detail.

Temperature control is important for some operations. Vinyl floor tile will not adhere well unless the building temperature has been maintained at a consistent level for a day or so with the tile in the building.

Plumbing is another highly skilled job and a lead plumber or layout man must have ability and experience. Many plumbing jobs look very simple, but today the work involves all kinds of problems with new codes and products. Inspections one never heard of a few years ago are becoming standard in most areas. For instance, a water service pipe from the water main to the house cannot come closer than 10 feet to the septic system or leech bed, in one widely used code. The reason for this is that a negative pressure in the water system could cause septic effluent to be drawn into the water pipe, thus contaminating the water supply in the main, and the house.

With the same thought in mind, all water outlets which can have hoses fastened to them, and dishwashers, require air-breakers to keep the water in the hose or dishwasher from being sucked into the house pipes if the water pressure reverses, or siphons. The air-breakers must be sweated on with solder to prevent their removal by children or others. A code book says air-breakers are required, but says nothing about soldering them in place, yet the inspector checks to see that this is done. This suggests how complicated plumbing is, and why a home builder needs a plumbing subcontractor to do the job.

1

The professional subcontractor knows from experience what is needed to complete the job. Time is money. Many do-it-yourself home builders tend to do the job too well, believing that if a three-bag cement mix is GOOD, five bag is BETTER. Too much time spent on rough grading also is a waste.

Are you using the best (not costliest, not cheapest) materials for your aims, life style, expectations and pocketbook?

With the multiplicity of new products and new versions of old standbys, even a general contractor may not keep up with all developments; but he will necessarily be more familiar with the quality and other specs than the first-time builder.

As each job is discussed in the following chapters, the text describes what a professional builder or building superintendent does as he schedules the men for the work and the decisions he makes.

SUB-CONTRACTORS CONTRACT

Sub-Contractor: _____

Address _____

Builder: _____

Date _____ Plan No. _____

W. Comp. Ins. Co. & Agent _____

Certificate No.: _____ Expiration Date : _____

Location of Work: _____

Total Price per House: _____ ($......................) Dollars

Terms of Payment: _____

Work to Be Performed and Materials To Be Supplied:

FOR SUBCONTRACTOR: _____
                        Signature and Title

FOR BUILDER _____
                        Signature and Title

2

LABOR AND MATERIAL

One of the most difficult jobs in costing is drawing the line between labor and material. What part of the house is labor and what part is material? In a sense, the whole house is 'labor'. For instance, lumber is labor because someone planted the tree, labor cut it down, labor loaded it on trucks, labor sawed it into boards, and so on until the carpenter nails it on the house. All cost is labor and when labor costs go up, costs in general go up. However, when considering what part of a house is labor and what part material, most people have in mind on-site work and delivered-to-the-job material. But even here one cannot answer the question unless the line between labor and type of material is drawn exactly, with a list of specifications. For instance, there is less carpenter effort on the job when factory-made trusses are used than with traditional roof rafters.

Today we tend to do everything possible in the shop, with shop-trained help, and to use less on-site skilled manpower. Shop work is less expensive, because the operations can be broken down to simple steps, each man being responsible for nailing one part or sawing one length of lumber. The best example of shop production is the modular or factory-built home. Yet as efficient as factory building can be, all problems are not eliminated. Shipping, for example, becomes difficult and costly.

The factory-made house is here and it often comes complete with furniture, drapes, appliances and everything connected up and ready for residence in one to three days on site time. However, zoning acceptance by the community, costs, quality, and many other problems have to be worked out, before factory-built homes take over. As of today, the savings, if any, are small.

In building your home, you can save money by using as many factory-built components as possible. There are trusses to replace roof rafters, for example. The triangular bracing and quality control in manufacturing these trusses make them a bargain. They are stronger than the conventional 2x6 or larger roof rafter. A truss is so rigid it won't give with two or three men standing on it, while ceiling joists 12 feet long and of 2x6 will deflect a half-inch when one man puts all his weight on top. Occasionally, after a house is built, a 2x6 will twist under attic heat and pull loose from the ceiling. Without a doubt, the truss construction is best, but meets resistance from old-timers and some building departments because they don't understand how it can be lighter and less expensive but better at the same time.

3

Prehung doors save skilled labor and offer a complete package — door jamb, hinges, lockset, everything at quite a saving. Even the trim is supplied, with mitered corners that have a metal spline and glue in them, so that they are superior to anything that could be done on the job. Some of these doors come prefinished with a coat of hard lacquer. All the carpenter does is put the two parts together and use eight nails. In ordering, care must be taken to get the right swing, since doors come hung to swing at right or left, and there are always several choices of door styles and lock sets. Decide on whose doors you will use before roughing the house because various fabricators require different rough openings in which to place the door assembly.

Trip time becomes important since a door, surprisingly, has some 37 parts, depending on the maker. If the package door is not used, someone must gather together the 37 parts and be sure that they are there when the carpenter is ready to install the door. If one part is missing, someone has to get it while the installer is still there. Otherwise he must come back to finish the job and be paid for two calls.

Kitchen cabinets are now factory built and vary from box-like cheapies to high-style furniture. Costs vary accordingly. Very seldom are cabinets built on the job. Few trim-carpenters can compete with the quality, styling and cost of shop-built cabinets. Also, the finish of cabinets is usually better when done under shop conditions with spray guns and dust control.

Component wall, floor and roof sections are available for on-site assembly. So are panels of pressure treated lumber for foundations. In some parts of the country such devices pay, but in others they do not.

The difference is in productivity, as against shipping costs. Modern on-site construction with machines is so productive that it is easy to compete with shop-built houses, especially when one adds the shipping costs from factory to job site.

Some other problems in factory built houses are the shipping damages. Very seldom can half a house or a whole large fragile package be shipped without damage. The longer and rougher the road the greater the damage potential. Then there is the factory built house delivered with everything but a certain size window or another essential part missing.

In todays headaches of shortages, a part can be missing for months, back-ordered, everyone working to get it, but it just is not available.

3

MACHINES

Machines are an important cost-saving factor, and in high labor-cost areas, sub-contractors should use every available type of machine to cut down on time and effort.

Our crews are equipped with portable power saws. This is a must. No one, not even an amateur, uses handsaws to rough a house any more.

There are 3,500 watt gas-powered generators, which give the crew self-contained independence. No matter where they go, they are not dependent on pole power or electric lines. An electrical unit like this will furnish power for four saws cutting at the same time. A gas-driven air compressor runs a pneumatic stapler which will drive seven-penny screw or ring nails that will break off before they can be pulled out. This system makes nailing effortless and fast. One man using the machine nails down a 1,000 square foot roof in two hours, as against four men doing the same for two hours by hand-nailing. The operator literally runs up and down the rafters machine-nailing roof boards or plywood in place.

Wall board crews carry taping machines which cut down on time and labor. Temporary heaters are important and can save a few key days in cold weather. Staplers, especially hammer staplers, save time on insulation and roofing.

Roofers with gasoline motor-powered elevators attached to aluminum ladders exert a minimum amount of effort to get the heavy bundles of shingles up on the roof.

Electric pipe-threaders and power drills help plumbers save hand-labor, and take the hard part out of this work.

Gas and electric-powered pumps will quickly drain water from a basement, footer holes, or crawl spaces, allowing work to continue.

Crews thus equipped are operating at maximum efficiency, are making money themselves, and are saving time and money for those for whom they work. This is especially true in high labor cost areas.

Trucks should be equipped with snow tires, jacks and a tow chain in case of bad weather. This can make the difference between keeping up a schedule or losing a day or two of work.

Other helpers are air hammers for poured walls; cement mixers for brick and block work; seamless gutter forming machines; and wall cranes for carpenters. As you near completion a telephone can be useful.

4

LAND ACQUISITION

Choosing and obtaining the land on which to build is a very important procedure. In order to avoid any serious problems in building, it is best to obtain a purchase agreement option to buy on the selected site for, say, a $200 deposit. Then, in effect, build the house and exercise the option. By that is meant, pay the balance due, and pick up the warranty deed. It has been the writers experience that this really works, and saves a lot of headaches.

The above procedure isn't quite that simple. To elaborate step by step:

1. First a deposit is put on the desired property. This amount must be negotiated but should initially be under $200 for a $4,000 to $30,000 lot.

2. In two or three working days an offer to purchase is drawn up with a larger deposit required. Again, this is negotiated in terms of the value of the property and the custom of the real estate people handling the deal. The purchase offer should require a warranty deed to be drawn up and held in escrow for a time long enough to get started building — say a three-month period. After this time the total amount for the lot is due and the deed delivered to the buyer from the escrow agent. The escrow holder is not the seller or anyone connected with the seller. This disinterested party (bank, title company, or other third party) holds the warranty deed for the specified time. Then he delivers it to the buyer when the seller has received the balance of funds at the end of the three-month period.

3. In this three-month period the house is well on its way. Barring bad weather, strikes or other holdups the professional builder could complete it in this period.

4. Even if the job is not completed, at least it should be past the stage where there is any doubt about the completion of the house. In this case, the option is exercised, that is the balance owing paid.

5. If any problems develop which would prevent building, your deposit should be cheerfully refunded. This situation will clarify itself before three months are up.

Two examples of the writers experience best illustrate the necessity of following this plan. A young couple came to me with the Offer to Purchase already executed, and a written sanitary test issued by the county on the property indicating that soil percolation was satisfactory for a three-bedroom house with one bath that they wanted built on this lot. They selected the plans, signed the contracts, and in a matter of days we applied for a building permit. After a week's delay (a holiday intervened), the building department of the local community informed me that the building permit was not approved because sewers and water weren't available. At first I though this meant that the sewer was across the street or down the block, as this was in a relatively small subdivision and a 100 x 300 ft. lot, not acreage. If that had been the case, it would have been easy to cut the street and extend the sewer or water in the same trench, or even bring it up from down the block, with each lot paying its share. However, the sewer was a mile away and the community estimated it would be two years before it reached the lot. As it turned out, the buyers received a refund on their deposit and went on to another lot in a different community. If they had bought the lot outright and received a warranty deed the sale would have been final and they could still have been waiting for sewer and water.

Another example was a lot in a big city on an unpaved street. Down the street, the writer had just finished a nice ranch home. This lot looked good and was slightly underpriced, but not enough to indicate any problems. It was a city lot with houses on either side indicating that the property was suitable for houses. A small deposit was given to the owner's real estate company to draw up an offer to purchase.

The lot front size was 50 feet, with the sewer location under the unpaved street in the center, and 20 feet deep. This is quite deep for a residential sewer but not a big problem, and possibly the reason for the good price. The excavator would have to be reminded that he needed a large machine to handle the depth. A small machine usually is used on narrow lots between houses for maneuverability. With this in mind, we called an excavator to explain it to him, advising him of the lot number, sewer depth, size of sewer, tap location and the nearest manhole.

It would be necessary to haul 45 truckloads of dirt away from the job since the lot was level and the proposed basement deep. The excavator has a job getting rid of dirt in a big city and may sell it or give it away if it is sand. If it is clay, it expands when dug, with 5 yards of hard clay expanding into 10 yards loose. Because this doubles the trucking volume, everyone is vitally interested in the type of dirt. The excavator quickly, by phone, contacted the original sewer contractor and found that this lot was fine yellow sand all the way down. So the excess dirt that would need to be hauled away would not cost anything because it was sand.

However, the trouble is that sand at that depth is under pressure and will flow like water. A special casing or hood is required to dig and tap the sewer in this situation. The work involves a minimum of $2,000 in extra cost. Since the lot itself was worth only $2,100, the situation was obviously impossible from a financial standpoint. Houses on each side were on septic tanks and fields not connected to the sewer, a procedure outlawed long ago by the city, but practiced when these houses were built. A request for the return of our deposit was made. At this point, the agent had prepared the offer to purchase and had done quite a bit of work so that the deposit return was strictly up to him because there were no restrictive clauses in the offer to purchase. But he did return the deposit.

Here again, imagine what would have happened if a warranty deed had been obtained with full payment and final closing. Further, the sand-versus-clay problem is not the only possibility. There are others such as zoning problems, structural complications, underground obstacles and potential title difficulties.

There is really no way to foresee these hidden problems. You simply have to smoke them out. In many areas, offers to purchase usually include a septic tank clause or a clause that says the offer is subject to the land supporting a septic system. This is a critical point that could prevent building and should be covered in the offer.

Electronic or pump septic systems are available and will soon be priced down to the homeowner level. They are now practical for commercial buildings, hotels, motels, offices, and restaurants, and may well solve a major land-use problem for residential, once the price is reduced through improved production and development.

To protect your interest in making out an offer to purchase, indicate at this time that you are only interested in the property to build a house on it, and the offer should include a clause to the effect that the deal will be completed only if a house can be built.

OFFER TO PURCHASE REAL ESTATE

THE UNDERSIGNED hereby offers and agrees to purchase the following land situated in the ⎧ City ⎫ Township of ⎩ Village

_____ County, Michigan, described as follows:

_____

_____

_____

_____

_____

_____

_____

ADDITIONAL CONDITIONS, if any: subject to the proposed house being built on this property. Included as part of _____ this agreement, drawings and specifications attached hereto. _____

_____

_____

_____

_____

IN PRESENCE OF:

_____ _____ L.S.

_____ _____ L.S.

_____ *Purchaser*

Address_____

Dated_____ Phone_____

8

5

FINANCING

Financing a new home can be the biggest problem of all for the do-it-yourself builder. Banks, savings and loan associations and insurance companies prefer to deal with reputable builders. A builder with 10 years' history in business and a gold plated credit report will get a better loan arrangement than any other type of applicant. Of course, the best type of financing is cash, but this doesn't happen often. Loans insured or guaranteed by the Federal Housing Administration or the Veterans Administration and numerous government-subsidized loan programs are more for professional builders than individuals, and usually involve a group approach in that five or more houses must be submitted for approval at one time. However, once a home is completed and a year old, the FHA will consider a mortgage commitment upon it. The rules for this change often.

The first thing one does in seeking a mortgage is to make an application. Some companies have their own forms, but in general these cover name, address, city, state, social security number, age, information on employment, assets, liabilities, work history and a list of past performance on loan payments.

Lenders have a standard routine set up by whoever authorizes them, the state or federal government. These require that the signed application from the people requesting the loan be kept on file. Also in the file will be a written credit account furnished by a recognized approved credit organization. The file also requires confirmed bank balances and a three-to-five-year history of credit references.

A credit report is further checked by asking the people their credit history, then calling banks, stores, and finance companies to confirm the references. Also, many credit companies have file card histories on millions of people. These serve as a check against damaging records that people will not voluntarily disclose, and will show if the applicant is able to pay for the loan or has a history of not being able to keep up promised payments.

Again, the loan company is authorized to loan only a certain percentage of the total appraised value of the whole package to obtain the maximum loan. The package must include specifications on everything — house, contents, well, septic system if any, outside walks, the land on which the house is to be built, and any improvements on the land. An appraiser considers all this, and estimates its worth at, say, $30,000. If the lender is authorized to go 66 percent, the loan would be approved for $20,000.

The percentage varies from 50 to 95 on residentual properties and 50 to 110 on apartments and commercial projects. (Apartments and commercial loans sometimes include 10 percent for developing and putting the package together.) The commitment amount is balanced between the ability to pay and the crisis sale value of the property. Of course, all of these percentages depend on the money market, and what the appraiser thinks, or whether the bank or loan company has or hasn't funds to lend.

Examples of two extremes on loans are: A young married doctor just out of internship gets a 95 percent loan on his new home. The loan company realizes that the man's skills will enable him to have a tremendous earning escalation and that he will probably have no money problems. Opposed to this: A married man with two older children who has a better-than-average job as a transmission mechanic. His record shows that he pays, but buys so much on time, owes so much and is often late paying, to the extent that the loan company submits a 50 percent commitment, or half of what is expected, because it does not really want to get involved without adequate protection.

Once a committment is made, a written form gives the terms and conditions. Term length on a mortgage varies, but 25 years is fairly standard. Monthly payments are leveled out over the whole period to amortize the loan, and the taxes and insurance are spread out over the year, one-twelfth for each month.

FIGURE YOUR OWN PAYMENT
PRINCIPAL AND INTEREST

	9%					10%					11%				
	5 Yrs.	10 Yrs.	20 Yrs.	25 Yrs.	30 Yrs.	5 Yrs.	10 Yrs.	20 Yrs.	25 Yrs.	30 Yrs.	5 Yrs.	10 Yrs.	20 Yrs.	25 Yrs.	30 Yrs.
$ 100	2.08	1.27	.90	.84	.81	2.13	1.33	.97	.91	.88	2.18	1.38	1.04	.99	.96
200	4.16	2.54	1.80	1.68	1.61	4.25	2.65	1.94	1.82	1.76	4.35	2.76	2.07	1.97	1.91
300	6.23	3.81	2.70	2.52	2.42	6.38	3.97	2.90	2.73	2.64	6.53	4.14	3.10	2.95	2.86
400	8.31	5.07	3.60	3.36	3.22	8.50	5.29	3.87	3.64	3.52	8.70	5.52	4.13	3.93	3.81
500	10.38	6.34	4.50	4.20	4.03	10.63	6.61	4.83	4.55	4.39	10.88	6.89	5.17	4.91	4.77
600	12.46	7.61	5.40	5.04	4.83	12.75	7.93	5.80	5.46	5.27	13.05	8.27	6.20	5.89	5.72
700	14.54	8.87	6.30	5.88	5.64	14.88	9.26	6.76	6.37	6.15	15.22	9.65	7.23	6.87	6.67
800	16.61	10.14	7.20	6.72	6.44	17.00	10.58	7.73	7.27	7.03	17.40	11.03	8.26	7.85	7.62
900	18.69	11.41	8.10	7.56	7.25	19.13	11.90	8.69	8.18	7.90	19.57	12.40	9.29	8.83	8.58
1,000	20.76	12.67	9.00	8.40	8.05	21.25	13.22	9.66	9.09	8.78	21.75	13.78	10.33	9.81	9.53
2,000	41.52	29.34	18.00	16.79	16.10	42.50	26.44	19.31	18.18	17.56	43.49	27.56	20.65	19.61	19.05
3,000	62.28	38.01	27.00	25.18	24.14	63.75	39.65	28.96	27.27	26.33	65.23	41.33	30.97	29.41	28.57
4,000	83.04	50.68	35.99	33.57	32.19	84.99	52.87	38.61	36.35	35.11	86.97	55.11	41.29	39.21	38.10
5,000	103.80	63.34	44.99	41.90	40.24	106.24	66.08	48.26	45.44	43.88	108.72	68.88	51.61	49.01	47.62
6,000	124.56	76.01	53.99	50.36	48.28	127.49	79.30	57.91	54.53	52.88	130.46	82.66	61.94	58.81	57.14
7,000	145.31	88.68	62.99	58.75	56.33	148.73	92.51	67.56	63.61	61.44	152.20	96.43	72.26	68.61	66.67
8,000	166.07	101.35	71.98	67.14	64.37	169.98	105.73	77.21	72.70	70.21	173.94	110.21	82.58	78.41	76.19
9,000	186.83	114.01	80.98	75.53	72.42	191.23	118.94	86.86	81.79	78.99	195.69	123.98	92.90	88.22	85.71
10,000	207.59	126.68	89.98	83.92	80.47	212.48	132.16	96.51	90.88	87.76	217.43	137.76	103.22	98.81	95.24
20,000	415.17	253.36	179.95	167.84	160.93	424.95	264.31	193.01	181.75	175.52	434.85	275.51	206.44	196.03	190.47
30,000	622.76	380.03	269.92	251.76	241.39	637.42	396.46	289.51	272.62	263.28	652.28	413.26	309.66	294.04	285.70
40,000	830.34	506.71	359.90	335.68	321.85	849.89	528.61	386.01	363.49	351.03	869.70	551.01	412.88	392.05	380.93

EQUAL MONTHLY PAYMENTS OF PRINCIPAL AND INTEREST TO AMORTIZE A LOAN OF $1,000

RATE ◊	9¾	10	10¼	10½	10¾	11	11¼	11½	11¾	12	12¼	12½	12¾
YEARS ◊													
10	13.08	13.22	13.36	13.50	13.64	13.78	13.92	14.06	14.21	14.35	14.50	14.64	14.79
15	10.60	10.75	10.90	11.06	11.21	11.37	11.53	11.69	11.85	12.01	12.17	12.33	12.49
20	9.49	9.66	9.82	9.99	10.16	10.33	10.50	10.67	10.84	11.02	11.19	11.37	11.54
25	8.92	9.09	9.27	9.45	9.63	9.81	9.99	10.17	10.35	10.54	10.72	10.91	11.10
30	8.60	8.78	8.97	9.15	9.34	9.53	9.72	9.91	10.10	10.29	10.48	10.68	10.87

EQUAL MONTHLY PAYMENTS OF PRINCIPAL AND INTEREST TO AMORTIZE A LOAN OF $1,000

RATE ◊	13	13¼	13½	13¾	14	14¼	14½	14¾	15	15¼	15½	15¾	16
YEARS ◊													
10	14.94	15.08	15.23	15.38	15.53	15.68	15.83	15.99	16.14	16.29	16.45	16.60	16.76
15	12.66	12.82	12.99	13.15	13.32	13.49	13.66	13.83	14.00	14.17	14.34	14.52	14.69
20	11.72	11.90	12.08	12.26	12.44	12.62	12.80	12.99	13.17	13.36	13.54	13.73	13.92
25	11.28	11.47	11.66	11.85	12.04	12.23	12.43	12.62	12.81	13.01	13.20	13.40	13.59
30	11.07	11.26	11.46	11.66	11.85	12.05	12.25	12.45	12.65	12.85	13.05	13.25	13.45

CONVENTIONAL MORTGAGE APPLICATION

Date: _____
Dept: _____
Branch: _____
By: _____
Ext: _____

No.

Husband	Property Address
Wife	
Street	**AMOUNT & TERMS REQUESTED**

AMOUNT & TERMS REQUESTED

Service Charge	Mtg. amt.	Int. rate	No. of months	Monthly Payt. prin. & int.
$	$	%		$

City	State	Zip Code

Phone No. _____ Bus. Phone _____

			Utilities Public Comm. Individual			Type of Heating
☐ Wood siding	___ Stories	___ Bedrooms	☐ Store Rm.			
☐ Wood shingle	☐ Split level	___ Liv. room	☐ Util Rm.			Type of Paving
☐ Asb. shingle		___ Din. room	☐ Garage	Water ☐ ☐ ☐		
☐ Fiber board	___ % Basement	___ Kitchen	☐ Carport	Gas ☐ ☐ ☐		
☐ Brick or stone	☐ Slab on ground	___ Fam. Rms.	___ No. cars	Elect. ☐ ☐ ☐		☐ Curb & Gutter
☐ Stucco or c. blk.	☐ Crawl space	___ No. Rms.	☐ Built-in	Sanit. Sept. Cess tank pool		☐ Sidewalk
☐ Comb. types	___ % Non-resid.	___ Baths	☐ Attached	Sewer ☐ ☐ ☐ ☐		☐ Storm Sewer
☐	___ Living Units	___ ½ Baths	☐ Detached			

LOT DIMENSIONS: ___ Ft. X ___ Ft. = ___ Sq. Ft.

GENERAL LOCATION: _____

Year Built		
ANN. R. EST. TAXES $	**ANN. FIRE INS. $**	**SALE PRICE $**

Occupation of Applicant _____

Legal Description of Property _____
_____
_____
_____
_____

Occupant _____ Phone No. _____

Appraiser can gain access as follows: _____
_____

What is your interest in this property?   ☐ Titleholder   ☐ Purchasing on Land Contract   ☐ Purchase Agreement Attached

Seller _____ Phone No. _____

Broker _____ Phone No. _____

COSTS:	NEW CONSTRUCTION		EXISTING CONSTRUCTION
Lot Purchased ___ 19 ___ Price Paid $___		Date Purchased ___	
Cost of Building $___		Purchase Price $___	
TOTAL $___			

(FOR BANK USE ONLY)

Appraisal Requested By: _____ Date _____
Signature

RECOMMEND	APPROVAL

☐ Pmts. 1/12 tax, fire ins., & FHA ins.

☐ Pmts. 1/12 tax

☐ Plans & Specs. and Certificate of Occupancy

☐ Final inspection

☐ Advance when enclosed

☐ Completion of repairs _____

Loan Percentage	Service Charge			
	Mtg. amt.	Int. rate	No. of months	Monthly Payt. prin. & int.
	$	%		$

☐ Special conditions _____
_____
_____
_____
_____

BY _____ Date _____

PERSONAL HISTORY

For what purpose is money to be used? _____

Applicant's Dependents: No. _____ Age _____ Age _____ Age _____ Age _____ Age _____

Age of Husband _____ Age of Wife _____ No. of years married _____

Amount of Life Insurance in force $ _____

Name of Employer _____

Address of Employer _____ Phone No. _____

Employer's Business _____ Position with Employer _____

Name and Title of Supervisor _____ No. of Years _____

Wifes Employment _____ No. of Years _____

PERSONAL FINANCIAL HISTORY
(Combined statements of Applicants, including contributions by members of family)

PROPERTY	INCOME

Bank Account _____ $_____
(Name of Bank)

Bank Account _____ $_____
(Name of Bank)

_____ $_____

Other Savings _____ $_____
(Name of Depository)

Securities owned _____ $_____
Other Investments _____ $_____
Personal Property _____ $_____
Automobile:

Make	Model	Year

_____ $_____
_____ $_____

Real Estate

Deposit on Subject property _____ $_____
Other Real Estate owned _____ $_____
_____ $_____
_____ $_____
Other Assets _____ $_____
_____ $_____
_____ $_____

ANNUAL INCOME
Salary for year . . . . $_____
Income from bus. . . . . $_____
Income from Investments:
   (a) Dividends . . . . $_____
   (b) Interest . . . . $_____
   (c) Rents . . . . $_____

Income from other sources:
   (a) Wife $_____
   (b) $_____
   (c) $_____

Total gross income for year

$_____

CURRENT LIABILITIES   Monthly Payt. Unpd. Bal.

Mortgage payment or rent _____ $_____ $_____
Automobile _____ $_____ $_____
Automobile _____ $_____ $_____
Debts, other Real Estate _____ $_____ $_____
Life Ins. loans _____ $_____ $_____
Notes payable _____ $_____ $_____
_____ $_____ $_____
Retail accounts _____ $_____ $_____
_____ $_____ $_____
_____ $_____ $_____

TOTAL $_____ $_____

NET WORTH . . . . $_____

I, (We) agree to furnish a title insurance policy and such other documents as may be necessary to enable your attorneys to examine the title, and to pay all expenses incurred.

I, (We) certify that all statements made in this application are true in every respect.

Accompanying this application is $ _____ for appraisal of property which is not refunded if appraisal is made.

Signed . . . .

Signed . . . .

Applicant is to call what credit bureau: _____
Remarks: _____
_____
_____
_____
_____
_____

A draw or construction loan type of mortgage would have a commitment form that states how many draws and how much for each one, plus the other information. This means that the builder puts the basement in, and then is entitled to some money from the mortgage company to cover the cost of the basement. A second draw would come after the roof is on, a third when finish carpentry work is completed, and the balance on the final inspection. Money lenders will not bother with this type of loan if money is in short supply, because it is not as profitable as a completion loan where the house is already up and requires only one inspection. Draw types require preclosing and quite a lot more servicing on the lender's part. Also the time element is against this program, as interest is charged on money drawn, and if the job takes too long the interest cost will cut into any possible savings. This is why scheduling and rapid completion saves money. People building their own homes sometimes find that problems cause costly delays, and even abandonment of the whole job. In that case, the lender must take over with a builder and finish the job to protect his investment.

On the brighter side, with a paid-for lot and construction draws, the house can be moved right along. An established builder has enough money that no matter what goes wrong he can keep up the schedule and perform in a short time to quicken the financing contract. Lenders are aware of this and, as pointed out before, will give a professional builder a better financing arrangement than a do-it-yourself planner. The lender checks the builder's track record carefully, paying particular attention to how strong he is financially.

Money rates must be discussed here too. The rate of interest in a certain large part of the country is described as the going rate. As an example, at this time in Northern Canada it is 11 percent, on the West Coast of the United States 9 percent, in the Central States 6 percent, in the Bahama Islands and the Carribean, 10. Now if this area's going rate is, say 9 percent, everyone says 9 percent. However, for a thin loan where the down payment is minimum and the customer scarcely qualifies, the lender might want 9½ percent. On the other hand, for a minimum loan with a well-qualified borrower it may be 8½ percent.

Interest rates on a mortgage in a certain area, with so much down, are thus set at a fairly constant figure. If someone comes along and offers you a mortgage at one percent lower than the going rate, you can be sure it is going to be made up some other way.

One way this special rate is offered is by an insurance company tied in to a combination mortgage with a lower rate and high-rate life insurance. Life insurance premiums make up the deficiency in the loan rate. The two are tied together in one monthly payment, so that if problems arise the borrower cannot drop the life insurance to reduce his payment. In other words, he must make full payment every month, or face forclosure. Upon checking, it may be possible to get the mortgage at better rates with the insurance separate, so that the two combined costs are less than the package deal.

Another way these special interest rates are obtained is to pick up points, or discount the mortgage, on closing. Thus, a 2 percent lower interest rate would be covered by a large premium payment on the closing of the mortgage or the loan would be discounted. This means that the mortgage debt would be, say $20,000, but the payoff to the home owner would be only $18,000, and the $2,000 would make up the interest difference over 25 years. A builder would have to handle this by adding $2,000 to the price of the house or go broke.

There are mortgage correspondents who can, by law, charge a finder's fee for getting a mortgage below the going rate and, of course, part of their fee makes up the difference in the interest rate.

For a last example, a building project may have houses selling with an interest rate of possibly 2 percent below the going rate. Consider the cost paid to obtain the lower rate and you will realize that the interest rate premium simply is added to the cost of the house. So the low interest rate becomes a bigger feature of the sale of the house than the cost of the whole package – house, lot, and mortgage. Or, the financing rate is low the first two years, then increases later. The seller in this way can feature a modest monthly payment which will go up in later years as the interest rate increases and the buyer's ability to pay increases, if he is young and starting at the bottom of a payscale.

Mortgage interest rates on homes are a bargain, as compared to rates on a short term loan, where the interest will be between 15 percent and 35 percent. These are regular rates for cars, applicances, or any short term under $1,000 loan. This is not an outrageous amount, but is necessary because of the cost of processing. As an example a $500.00 loan at 6 percent per year comes to $35.00. This amount will not even cover the cost of taking the application, which involves: an office work station, people to do the interviewing, credit report, telephone calls, and all the other costs applicable. Thus, they must charge higher interest rates to make small loans pay, and be available.

A $15,000 mortgage at 8 percent is a different story. There is enough total interest to make the loan worth while at that rate. However, if it is not profitable the amount necessary to make it pay is added on as points, closing costs, service charges, or some other cost, but in no way does the interest cost approach that of the small loans.

Further, the true cost of interest is related to your income. If you itemize tax deductions, your interest is deductible. For round figures, someone in the 50 percent tax bracket paying a mortgage interest rate of 9 percent, the TRUE RATE OF INTEREST is 4½ percent, and less as the tax rate goes up. Another big deductible item could be the sales tax applied on materials when you are your own contractor.

YOUR LOT DOES NOT HAVE TO BE PAID FOR TO BUILD ON IT, under some financial plans. I first experienced this years ago when a young man came to me and requested that I build him a house on his lot which was not paid for. The figures he presented were, property on a beautiful river worth $24,000. He owed $5,000 and wanted an $11,000 new home built on it. He had a good job, good credit, but no money. The mortgage appraiser went out to see the location, turned in an appraisal of $24,000 on the land and indicated they were willing to lend the required $16,000 for the house and the balance on the land.

Today as a builder we simply lend the customer the money to pay off the land contract, then write a new land contract to cover our money, build the house, then the bank or lender re-mortgages the whole package. Of course, the borrower must be able to qualify for a loan large enough to take care of the house bill plus pay off our lien on the lot.

14

6

DISCOUNTS AND PRICE GUIDES

The word discount has meaning only when you ask, on what? Wholesale discounts involve buying large quantities. The more one buys, the less the price per unit. Suppliers have a breakdown of so much percentage off if the bill is paid between the first and 10th of the month after delivery. Two percent is the figure often quoted; however, on mason supplies it is 5 percent. This probably is due to the fact that mortar and sand are a small amount of money in terms of the over-all cost of a home.

The material handlers patronized by a do-it-yourself builder, and the workmen that subcontract, know they have a one-shot deal. A general contractor, on the other hand, is responsible for the entire job and its quality and often has other work in progress too, therefore, suppliers are liable to give an established builder a better discount and more service.

Discounts usually are offered as a sales incentive. A supplier says, "I'll give you 30 percent off the list price on lumber." This means absolutely nothing because you don't know what the list is. Wholesalers have their own lists. Various associations and dealers have different lists. As your own builder or subcontractor, what discount will you get? The supplier's profit on one individual sale must be, say, 10 percent. He sells to a small builder who builds 25 houses a year, but his cost of sale for the 25 houses is about the same as it is for the one-time sale he gets from the do-it-yourself home builder. No matter how one states it, the established builder buys for less money. Any discount earned by paying cash or buying material on sale will not be as great as the general contractor will be getting.

A good example is windows. If you go the the local window manufacturer or lumber company distributor you will be told that because you are building your own home you will be given a good discount because you are a good Joe, building your own home like that. The price you will get may be 40 percent off the retail price list, or $250 worth of windows for $150.

On the face of it this sounds like a good deal and it may well be, under these circumstances, where you are talking about only 11 to 15 windows and just one house. The man is selling just one order and he will have to check where it is to go, who will pay, will have to make an arrangement to check the plan and to figure in a bad debit risk, all for one small sale.

By contrast, the professional builder has one phone call to list the windows to be sent to the job. He knows down to the inch and the penny what he wants and needs. His cost is 40 percent of the retail list, or a 60 percent discount. With the same set of windows mentioned above, the builder's cost is $100 instead of $150 for the $250 set of windows at retail level. He knows he has to pay between the first and the 10th of the month after billing or he won't get this kind of discount, so he pays on time.

There are many tricks to the builder trade aside from this. Some builders have their own lumber companies and cabinet shops and buying groups which enable them to purchase franchised material at wholesale levels. Certain lock companies will only wholesale to lumber companies, and a broad spectrum of other producers will only sell to dealers. I know people who have business names suggesting they are dealers, but they aren't; they simply have the business identification to enable them to buy wholesale.

Grouped under discounts are a wide variety of charges that tend to offest discounts. Some of these are zone-shipping rates, boxing and crating costs, delivery charges, demurrage, towing charges, late charges and strapping costs. Such charges are added as labor costs grow — the price is the same, but the added charges make up the difference.

Such charging systems are competitive; everyone in the area does it and the one who doesn't cover his costs and profit in this way goes bankrupt.

If one orders supplies in the spring when the ground is soft and the access road is not adequate to support the truck, the risk is passed on to the builder if he wants the material delivered. This is done by stating that if the truck gets stuck the builder buyer will pay for the towing. Again, what this means is that the supplier is selling so close to costs that he can't afford to pay for having his truck pulled out of the mud.

All supplier truck drivers are highly skilled drivers, but there is a risk in the spring on some ground. Demurrage in the context mentioned here is money charged for overtime in unloading concrete. A half hour is allowed and if the unloading takes longer the truck and driver time are charged to you.

So how does one tell what the cost will be? It is best to use this kind of direct approach when dealing with a supplier: How much will it cost for a list of material delivered on the job within two days of ordering, with my check to you in the exact amount paid between the first and the 10th of the month after delivery? In this manner, you clearly state that all you are interested in is the total cost, without plusses for shipping and other things.

Most bills are presented before the first of the month with the request that they be paid between the first and tenth. Some companies charge a penalty if payments are received after the 10th, and a further one percent or more per month carrying charge on the unpaid balance. At the writing of this book, interim funds are running 10½ to 11 percent. This means that a subcontractor who wants to borrow or extend his line of credit to pay a bill on labor or material must go to the bank and borrow at this figure. Subs protect their credit in case the main contractor or builder does not pay them on time. Subcontractors often work on a very narrow margin, and need to be paid as soon as the work is completed. If the builder agrees to do this the subcontractor can give him a better price.

Thus price quotes should include a statement by the buyer as to how payments are to be made. Best prices will be quoted on assured quick payment.

16

7

PAYING ACCOUNTS

The best way to pay for work or supplies is immediately upon billing and when the work is done, or material delivered. Ideally, one should be able to sign the shipper and get a copy of it at delivery. This way the list of material can be checked to make sure everything is in order.

Subcontractors have various ways of collecting their money. The best for builder and sub is to fully pay upon completion and inspection. However, completion and inspection do not happen at the same time so a good plan is to hold back 10 percent until the job passes inspection. If there is not an inspection and you are satisfied with the work, pay the sub in full. The subcontractors have a fraternity of their own, a loyalty to each other, and you will find that the excavator if paid immediately will be your best reference for the men to follow. Keep in mind again that these men are working against tremendous labor costs. If a nail is sticking out and needs to be pounded in, do it yourself rather than call the man back and insist that he do every little thing. We do everything we can to keep subcontractors profits up. Any item that is wrong upon completion should be called to their attention immediately so that a minimum of time will be lost in correcting it. Payment on work done is often in two steps, rough electric and finish electric; and rough plumbing and finish plumbing. Subcontractors should present a bill and be paid by check. If material is a major part of the bill, as in poured concrete, the check can be made out to both the worker and the concrete company, thus making sure the material bill is paid.

Never pay anyone in cash as this makes record keeping very difficult. For any small item purchased at the hardware store, get a receipt, keep it in a handy place, then you will know exactly what your job costs. A certificate of insurance covering workmen's compensation should be in your hand or you should deduct it's cost from the bill if the sub doesn't have insurance. Ask your insurance man for the rate so that you can get the deductible insurance cost correct. As an example, carpenters pay $3 per $100 dollars for workmen's compensation. Cleanup men only pay half that amount.

Always insist on a bill before paying. It should have their company letterhead on it, or the sub contractor's name, state what it is for, and the job location. When you pay this bill record your check number opposite the amount paid on the bill and on the cost estimating sheet (page 22). This acts as an immediate cross-reference for accounting. Then keep the paid bill.

17

8

PLANS

Most housing design is planned to keep labor and material at a minimum. The greatest enclosure for the least amount of material is a circle. Next is a square. However, building materials in the United States are manufactured in 2-foot increments. Acceptable spans for floor beams or joists are established from testing and these tests set the standards for building codes. The most acceptable span for the floor is a 2x8x12 at 16" on center. This means with 4 inch bearings, one on the block side and one at the beam, the wooden member will support normal living. Two of these placed end-to-end with a lap at the beam will measure 24 ft. for a frame house. This sets the design size. To go greater than the code span requires a larger lumber with a larger cross section and to go shorter in less than two feet lengths requires labor to cut the piece off, and waste to throw it away. A 23 ft. house would cost more than a 24 ft. house, because of the added labor and waste.

Plans obtained from known sources take this into consideration but it is worth mentioning, because some do-it-yourself designers do not realize the effect of this 2 ft. standard on construction, which applies to walls, ceilings, brick, drywall, and every part of the building.

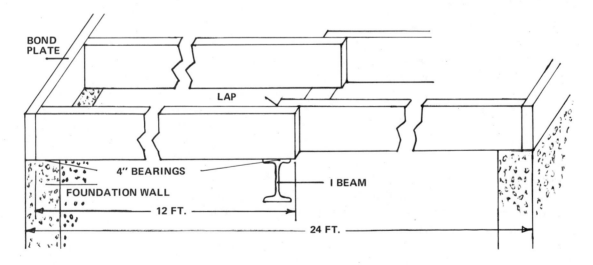

Starting the house involves plans or blueprints. These may be stock plans available from plan services through newspapers and magazines, or can be from a draftsman. Some lenders have a home owners monthly with a featured house plan. Any of these will do to help pick out a design close to what you would like, which then can be modified as to dimensions, detailing or, say, different windows. A draftsman can make these changes, or can make actual working drawings from the sketches presented in the paper. Usually the plan service will do this on request, for a fee. If you do plan changes, order only one set of plans. Changes can be made on these working drawings with a red pencil or felt marking pen. One building code will allow changes up to two feet without redrawing the plan; others require it redrawn for any change. If you plan no changes, buy four or more sets of the same plan.

Another way of obtaining plans is to find the house you like from builders' models, subdivision open houses, or a friend's home which is close to, or exactly what, you had in mind. Ask the builder or owner if he will sell you a set of plans. This has the advantage of a chance to check visually what the proposed home will be like. Most people cannot envision sizes and shapes from an architectural drawing. If the plan cannot be obtained, a draftsman or architect can, by looking at sketches or your friend's home, quite easily draw up something similar to it or just like it with the variations you would request.

Stock plans have the advantage of being economical, practical, and standard in sizes of contruction material. An added advantage is that they have been built many times so that mistakes have been corrected. People ask me, who designs our homes? I tell them people, our customers, design the houses.

As an example of this, an up-and-down flat or up and down apartment plan had a door between the kitchen and dining nook. After the units were built the occupants would take the swinging door off and store it in the basement. The swinging door would get in the way, and there was not enough room for a sliding door pocket in the wall, so the master tracing was changed to include an arch in place of the swinging door. This made the plan nearly perfect.

In small houses, about 1,000 square feet, every inch is important. Often people want to move a portion of the plan this way or that a foot or so. This is critical, not as simple as it sounds, and can cause unforseen troubles. Changing the location of a window by moving it over so it will look better on the outside can cause inside relationships to change. The move may effect heating, wiring, pipes or doors, and can cause other complications. Some codes require kitchen windows to be at least 3 feet from stove locations so that the draft does not blow out the flame. Moving a window here could cause a conflict with the code.

Then, of course, there is the architect who will draw a plan for any kind of house the customer wants. Not all architects do residential work, and when they do they usually are involved in more expensive homes or in multiple construction. But this kind of custom service is available if you look for it.

Relative size of a house is determined by the square feet of floor space, but square footage is not a good guage of the price. A 1,000 square foot house at $20 a square foot

MEASURING SQUARE FOOT SIZE

Same House — Different Methods of Measuring
Used by Various Codes

Square Foot ⌀

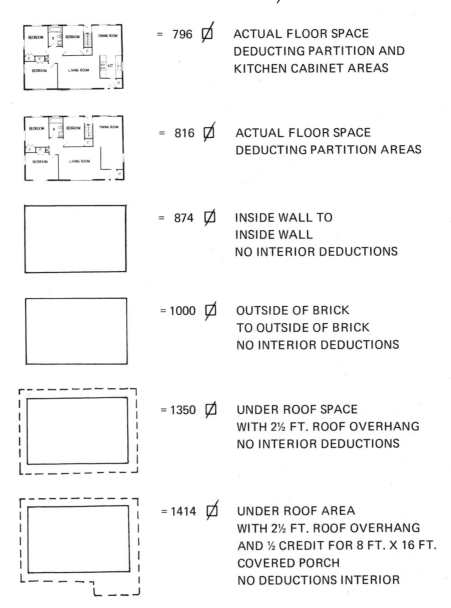

= 796 ⌀   ACTUAL FLOOR SPACE
DEDUCTING PARTITION AND
KITCHEN CABINET AREAS

= 816 ⌀   ACTUAL FLOOR SPACE
DEDUCTING PARTITION AREAS

= 874 ⌀   INSIDE WALL TO
INSIDE WALL
NO INTERIOR DEDUCTIONS

= 1000 ⌀   OUTSIDE OF BRICK
TO OUTSIDE OF BRICK
NO INTERIOR DEDUCTIONS

= 1350 ⌀   UNDER ROOF SPACE
WITH 2½ FT. ROOF OVERHANG
NO INTERIOR DEDUCTIONS

= 1414 ⌀   UNDER ROOF AREA
WITH 2½ FT. ROOF OVERHANG
AND ½ CREDIT FOR 8 FT. X 16 FT.
COVERED PORCH
NO DEDUCTIONS INTERIOR

20

would run $20,000. But the price could be more or less depending on what is included. So a square foot price can be considered only in general terms as a basis for comparison. Another way of relatively pricing a plan is to find a comparable house that is for sale, and subtract the estimated cost of the lot, well, and septic tank or sewer and water costs. This will give a fairly close figure. People often come to me with large complex plans from a plan service and are shocked that the rough estimate is twice what they thought the house would cost. Even if the comparative house pricing procedure is off a couple thousand dollars it is not half off, and indicates some kind of starting figure.

One reason price cannot be related to square footage is that the method of measuring a square foot is not standard. Usually one thinks of a 40 by 25 ft. building as being 1,000 square feet of house, but some building codes require inside wall measurements with partitions taken out of the total area. Other measuring codes will give half credit for attached garages or covered porches, and one code accepts measuring from the tip of the gutter to the tip of the opposite gutter. Thus a 2 foot overhang plus gutter adds almost five feet to the "size" of the house.

A square foot price can be distorted by a simple failure to state what is included. A bargain buy of 2,000 square feet under roof means just that — a roof and walls without plumbing, electric power, furnace, kitchen cabinets, trim, or bath fixtures. It all might just have a front and back door!

A close estimate of cost can be obtained by adding up all the latest figures on labor and material. Check this against the cost of a similar new house already built. It will give an idea of what the spread is between cost and sales price. However, in figuring the cost, add about $500 for theft, material lost to the weather, trucks running over supplies, broken windows, and many small items that come up unexpectedly and cannot be figured in advance. The latest estimate will be the closest, because inflation is a large factor in costs. Residential building prices have gone up at a rate of up to 12 percent a year since 1963, although the rate has slowed in 1972. Thus, estimates must be up to date to reflect a true picture. It is even a good idea to prorate or advance estimates because carpenter labor by the time the job is started will be 10 percent more than when your cost figures were estimated because of inflation.

Also, getting a signed bid price for an activity such as carpentry is good, but without proper checking and understanding the actual billing can be quite a bit higher than what you expected.

An example: if the lumber is delivered just off the street and not up against the foundation, the carpenter would want extra money for the extra work required to get it moved up to the basement.

The following page illustrates a cost estimating sheet, the same form we use to keep track of and record payments made to subs for each house built.

Estimate figures for subs can be written in the PAID column of this sheet in pencil then used as a reminder of the bid, lastly written over in ink for a final record when paid.

Permit No. _____

Street No. _____ Between _____

Bank or F.H.A. No. _____ Sub-Div. _____ Lot. No. _____

Job Begin _____ Finish _____

	PAID	Check No.	DATE		PAID	Check No.	DATE
Attorney				Lot			
Architect				Linoleum			
Abstract				Lumber, Rough			
				Lumber, Rough			
Bricks				Lumber, Finish			
Bricks				Lumber, Finish			
Bricks, Cleaning							
Bricks, Laying				Mirrors			
Blue Prints				Mortgage Application			
Blocks							
				Oil			
Cabinets, Kitchen				Oven, Stove			
Caulking							
Carpenter, Rough				Plastering			
Carpenter, Finish				Permits			
Carpenters, Finish				Painting			
Cement				Painting Repair			
Cement				Plumbing Permits			
Clean House				Plumbing, Rough			
Commission				Plumbing, Finish			
Deed Rec.				Roofing			
Dirt Haul							
Doors, Outside				Sand & Gravel			
Doors				Siding			
Doors				Stones			
Drywall				Shades			
				Steel I Beam			
Excavator				Steel Column			
Electric, Rough				Septic Tank			
Electric, Finish				Supplies			
Electric, Temporary				Survey			
Electric Bills				Storms, Screens			
				Tinning			
Floor				Tile			
Floor Laying				Taxes, City			
Floor Sanding				Taxes, City			
Formica				Taxes, County			
				Taxes, County			
Glazing				Tree Removing			
Glazing Repair				Trusses			
Glass							
Gov. Stamp				Window			
Grading				Weatherstripping			
Gas Connection				Wall Poured			
Gas Bills				Well			
				Miscellaneous			
Heating							
Hardware, Rough				Sold For			
Hardware, Finish				Extras			
				Deposit			
Insulation							
Insurance							
Insurance Claims							

9

PLOT PLAN

The plot plan is, in its simplest form, a rough, unscaled sketch of the position of the house on the lot or acreage. In the elaborate form it is drawn to scale, and shows elevations, bench marks, existing houses on each side, distances to the front and near corner elevations, all underground information, walks, drives or other proposed cement work, trees and landscaping. Community requirements vary. Some are simple and some are complex. So starting from the simple:

Side yards and setbacks are the factors that determine where the house goes in relation to the property lot line. They are described, for example, as five feet with a combination of 12 feet. This means that the minimum must be five feet, the other seven feet, or more. The five feet could be more, too. To go on, often the distance between houses is specified, such as twelve feet, thus, in the above case the five feet minimum would be seven feet if the house next door was the required five feet from the lot line.

The best of plot plans is called the topographic plot plan or topo for short. This gives the exact legal description, then a scale plan detailing steel or wood stakes outlining the lot. These are set in place by a surveyor as the first on-site physical activity on the lot. After these are set from the subdivision plat the surveyor makes a field sketch of the dimensions for location of everything on the construction site — other buildings near the property or next door, and roads or streets. The elevations are shown too for each corner of the property and the buildings adjacent. This is the distance up and down between a bench mark or something that is stable to use as a base or starting point, usually the center line of the road, a fire hydrant, telephone pole or the corner of a building. It is designated as 100.00 on the drawing. The ground is shot say one hundred feet back of the road where the house will be located. A shot or sighting is taken with a level or instrument called a transit. If the ground there is 98.00, this means that the soil at this point is two feet below the center line of the road. The field rough sketched information is taken to the office and these notes made into a drawing or tracing eventually ending up as a blue print. The draftsman gets his underground information from the local sewer and water department. Telephone, gas and electric locations must also be checked and put on the topo, which shows with notes the size and depth, and the tap or riser locations, for the sewer. Caution: an excavator who cuts or damages a telephone ground cable should realize that in some cases he is liable for the damage, and if it is a long line system the charges can add up very fast for even a brief distruption.

Now the draftsman will locate the house on the topo using the house plan for size. Then put in the elevation, this means the distance from the center of the street to the corner of the house, up and down. A 100-foot setback 2 feet above the road would be an ideal example, as the road very seldom floods, and a two foot rise for a 100 foot setback is hardly noticeable. This would appear on the topo as a fraction +2.00 over -2.00 meaning two feet above the road for the finish grade, and the present soil at this point is two feet below the road.

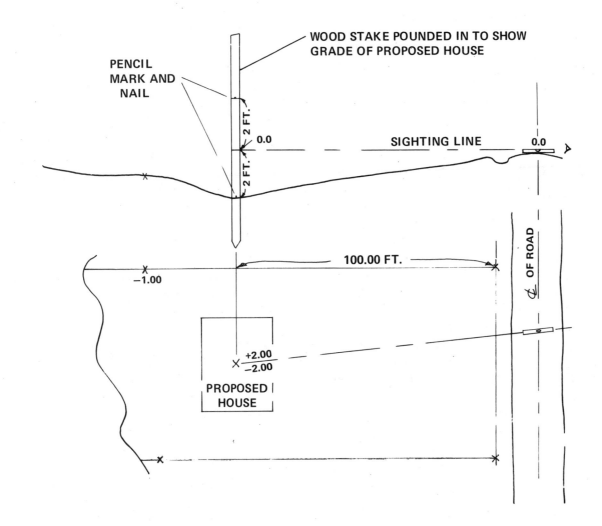

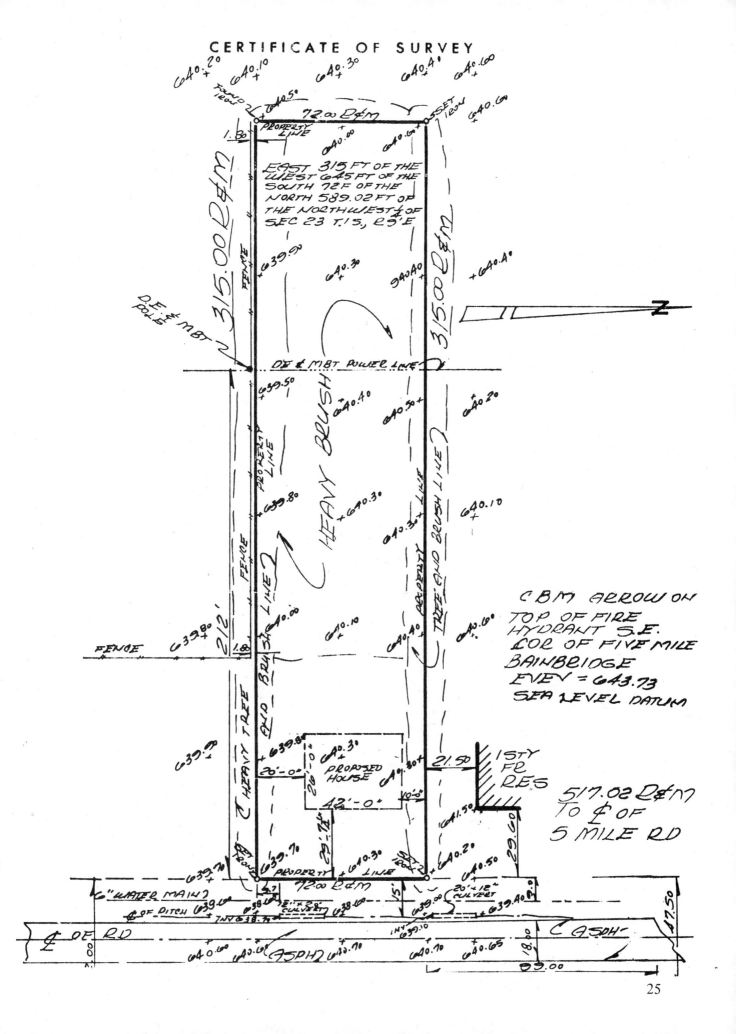

Six to ten copies of the finished topo will possibly be needed. Two for the FHA or financing company, one for the excavator, one for the basement man, two for the local building department and a couple for the file.

The cost of the topo is based on time consumed to do the work, and varies from $60 to $400 for an average lot. The latter occurs when there are no stakes, reference points or monuments for the surveyor to check, and he has to start from the beginning and check out the whole subdivision or section, or when the ground is extremely hilly, treed or otherwise obstructed.

Most mortgage companies require a final survey before disbursing funds. The topo or simpler line drawing of the lot by the surveyor, with the house located, will satisfy the requirement. It is important to realize this, otherwise, the mortgage company will order a mortgage survey which is simpler than the topo but involves a double cost and will be charged to the owner builder, or to the mortgage closing costs. The topo may have to be updated but this cost is very small. A 4x5-inch picture of the completed house is sometimes part of the mortgage survey or possibly more than one. Simple black and white photos will do, and anyone can take them. One front view of the whole house, one side picture of the house showing the street view, should do. Identify the pictures with house number, street, owner's name, and mortgage number.

The survey of the lot for a line drawing should be ordered before buying the property, as this will show all the underground and easements on acreage for pre-planned roads, pipe lines and situations of which even the seller is not aware. This can be the preliminary for the topo. Houses should not be built over easements. If a certified survey isn't required, all this information is obtained by looking at the county records. Someone handy can make a sketch or line drawing from these records, called plats.

Thus we see the possibility of doing and paying for this survey once, for the three uses. First buying the property, second plot plans or topo with which to build, and last a mortgage survey.

Another use for the layout is for the well and septic plan, if required by the community.

When choosing a surveyor, if the man has done work in the area before or worked on boundary properties he will quite frequently charge less than a company unfamiliar with the area. This is based on notes and check points he has developed in the previous work.

Surveyors use monuments, which are basically points set in the ground. Usually the areas are laid out in one-mile squares, over which roads are built, especially in rural areas. That is why when an east and west road crosses, one often observes a miniature manhole concealing an up-ended pipe on the crossing point of the two road centerlines. A steel bar is in the center of the up-ended pipe below the road level; this is a reference point, a monument. It is set deep into the ground surrounded by pavement and is about as permanent as it can get. Measurements are then taken from this reference point, or corner monument.

10

SCHEDULING

Order this previously described plot plan immediately, as this involves moving a survey crew into the area and much desk and phone research for sewer, water or well and septic requirements and placements.

The plan we follow to keep our jobs moving, which saves time and money, is that subcontractors who have the right price are told that the job will be ready a couple of weeks ahead of time. Then three days before the job is to be done we ask that they plan for it that day. If they cannot be on the lot in three days, we have two alternate subs for the same job to call. If they cannot go in, another type of job is scheduled. The night before the contractor is to work on the house, phone and confirm that he will be there. If he does not show up, give him three days to make it and then try to get someone else.

Many times a working crew does not show up because of extra work they did not plan for on the last job, or because of hot weather, cold weather, a machine breakdown, a shortage of material — who knows why, construction is a tough business! All of our regular men have reliable records and perform unless something drastic occurs. If it does, the whole schedule must be shifted back for all the crews who follow.

The Progress Sheet illustration shows a form which helps keep track of when a job was ordered and the completed date.

Probably the one main problem which will occur is the money problem. Time is money and if the project is not moved along rapidly by a constant telephone and personal contact with the subcontractors, the partially completed home will cost so much in interest that all advantages will be lost. If you have lots of money and are having fun changing and adding things as you go along, it will come out fine. Anyone can build a house if costs are not a problem, because no matter what happens, money and time can fix it.

The Progress Sheet illustrated shows a form which helps keep track of which sub follows which, when the job was ordered and the completion date. There is a possibility that two subs can work on the house at the same time, the critical question here is, if they do not get in each others way. A trim carpenter working in one end of the house may not disturb the tile setter doing a bath room in the other end.

Progress Sheet

Lot No. _____ House No. _____ Street _____

Between _____ and _____

Subdivision _____

City Permit No. _____ F. H. A. No. _____ Mortgage No. _____

Purchaser _____ Phone _____

Purchaser's Address _____

Plan No. _____ Elevation _____

INSPECTIONS

1st City _____	Crock to Iron _____
1st F. H. A. _____	Inside Drains _____
2nd City _____	Rough Plumbing _____
2nd F. H. A. _____	Copper Service _____
Final City _____	Rough Electrical _____
Final F. H. A. _____	Gas Service _____
	Gas Meter _____
	Electric Service _____
	Public Walks _____

EXTRAS

SUBCONTRACT	ORDERED	COMPLETED	MATERIAL	ORDERED	COMPLETED
Survey			Conductors and Trap		
Excavating			Block		
Block Layer			Beam		
Rough Carpentry			Rough Lumber		
Roofing			Stanchions		
Siding			Chimney Material		
Prime outside paint			Lath		
Rough Plumbing			Flooring		
Rough Heating			Trim		
Rough Wire			Inside Doors		
Chimney			Cupboard Doors		
Clothes Chute			Hardware		
Inside Drains			Switches and Plugs		
Insulation			Electric Fixtures		
Plastering					
Cement Work					
Floor Layer					
Heat Registers					
Tinning					
Furnace					
Finish Carpentry					
Tile Floor					
Finish Plumbing					
Bath Tile					
Weatherstrip and Caulk					
Painting			FINAL INSPECTION		
Sanding and Varnish					
Shoe and Hardware					
Linoleum					
House Cleaning					
Shades					
Glazing					

11

HEAT SHEET

The heat sheet is a part of the plan's display submitted to the building department with a request for a permit to build. It is not always required, and if necessary must be made up locally, in the area where the building is going to be put up. Heat loss is higher in a colder climate than in the southern latitudes, and the local heating engineer is acquainted with his area needs. These sheets are similar to the floor plan, showing size and location of registers with a note as to the rough opening sizes for the rough carpenter. Figures for the heat loss on each room appear in a note placed in the center of the room. A final total heat loss is figured and this is given to support the size of furnace required to heat the house. The furnace layout, duct work or hot water heat pipe runs are shown on a drawing of the furnace room, attic, crawl space, or basement, wherever the heating unit is located. Heat loss and furnace size take into consideration the amount of insulation in the building, ventilation, style of window, double glass or other, size of windows, door openings, and caulking and weather stripping. Heat sheets cost from $25 to $100 dollars for a three-bedroom home and require some time to prepare, for getting the figures together and checking with the community as to what is needed. Some communities want perimeter registers under each outside window. Thus all registers must be in the floor, fanning the heat up and away from the window, providing a hot air screen between the window and the room. Whether this is the best heating system is questionable. All types have advantages and disadvantages and various ways of being installed. Much of this is up to the local heating inspector as to how he wants it done. Sometimes one does not have any choice.

Heat sheets can also be planned to show central air conditioning layouts even if the air conditioning is to be put in later.

Keep in mind that once the heat sheet is approved every register or pipe must be installed just as the heat sheet is marked. Many times people tend to change the location of a heat outlet because a couch or piece of furniture will go over the outlet. If such a change is deemed necessary, it should go all the way back to the heating engineer, then be approved by the building inspector before making any change on the job. Even if the register location is only changed by one foot, the above procedure should be followed.

The reason for all this is that complaints go down through channels but the buck stops at the builder whether he is a do-it-yourself first-time builder or a lifetime professional.

Some one who lives in the house says it is cold, it will not heat. The building inspector checks the house over and finds the heat sheet and specifications were followed exactly; but if every register is not in the specified place someone will have to change it. This goes for furnace make and size too.

Years ago a house could have a cold back bedroom or other parts of the house. Today it is almost impossible to build a house with a heating problem. Pressure systems allow the furnace to be placed just about anywhere, and heat can be forced to every part of the house under enough pressure so that cold corners are impossible.

Humidifers should be mentioned here. In real cold climates heat tends to dry out the house excessively, requiring a humidifer to put the moisture back into the air and the house itself. This can also be done with household plants, open top fish aquariums and clothes hanging in bathrooms to dry out.

Excessive humidity can wreck a house and this is the problem with mechanical humidifiers. Home owners often do not turn them off even in the summer, when normal area humidity becomes a big problem. For this reason a lot of builders will not install humidifiers in geographic locations where winters and summers are extreme.

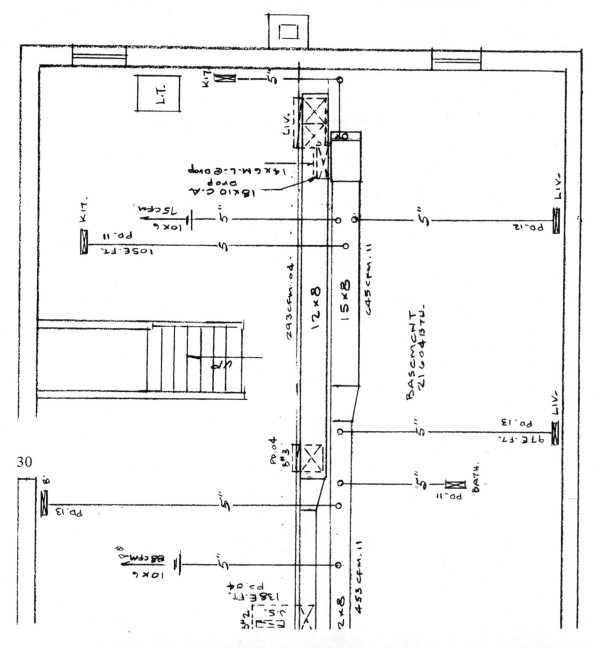

30

12

INSURANCE

In planning the construction of your home you will need insurance. As with labor, high costs apply to insurance too. Not so many years ago, insurance on a new house used to cover glass breakage with no deductible clauses. Today, few insurance companies will cover glass and most policies have deductible clauses in them. To glaze every window in an eleven window house used to cost $37. Today that amount won't even pay the travel time to the job and back for the glazer and truck.

To begin with, a policy is needed called a builder's risk. Some companies write a regular home owners policy with a builder's risk clause to cover the house while it is under construction. Builder's risk is a special type of policy to cover the house while it is under construction in case of fire, windstorm, and extended coverage damage. Vandalism is included in this coverage. The whole policy should have a deductible clause, to make the policy less expensive. It specifies nonoccupancy; that is, no one lives in the house, and it is not furnished. Upon completion, and the issuance of a certificate of occupancy, the normal home owner's policy is in order if a home owner lives in the house. If the home is rented, the proper policy is a fire and extended coverage policy. These lines of difference are exact, and are applied to the very word in case of a claim. All the insurance contracts must be perfect, and complied with exactly.

As to theft, this type of insurance is so expensive that no one can afford it. Theft is a big problem in building and it is increasing as costs increase on material. Law and order convictions are almost impossible to obtain. Prosecution is so difficult and time consuming that it is not worth while. One loses more in working time than the stolen article is worth, and you still may not get a conviction. A standard excuse for a caught thief is that another builder sent him over to move the lumber from your job to another and he made a mistake by taking the material from the wrong house. One of the best ways to handle theft is to get the job done in a hurry. This means large crews of carpenters — 10 or 12 men can do a small house in one day if well organized and accustomed to working together.

Fencing around a construction site, with a light, may pay off for security. Even a partial fence, a temporary snow fence, or 50 ft. of chain link stretched across the front lot line, makes a statement: "Keep Out". It shows the owner is aware that stealing is a problem, do not take anything.

31

"Dog" and "Call Police" signs will help cut down vandalism and theft. So will friendly neighbors who will help watch the construction.

Workmen's compensation insurance is the most troublesome type of coverage for the do-it-yourselfer. The successful builder who gets by without the insurance is just lucky, and is taking, or has taken, a tremendous risk that he doesn't understand. The risk is that if the workman is hurt on the job he can hold the owner of the property liable for his injury, or his estate can collect if he dies. Huge amounts of money can be involved.

Such claims can be far in excess of the value of the house and lot, or the average person's ability to pay in case of a serious accident. Regular builders handle this by requiring all subcontractor's groups to have their own workmen's compensation insurance. Then the builder takes out a blanket policy to cover anyone who does not have insurance for one reason or another. This policy also takes care of the day laborer that the builder uses only for a day, pays him a flat rate, and probably will never see again.

32

The mechanics of the procedure are quite simple. The builder posts a premium ahead of time, then at the end of the year, all the subcontractors furnish insurance certificates sent directly from the insurance company to the builder. The builder's books are audited to check if any labor payments were made to subs not covered by certificates. If no certificates are available to cover a sub's wages, the builder is charged so much for each dollar paid, and this is taken out of the original deposit. Then the builder deducts this premium from the subs pay if the sub says he had insurance, and does not. At this time the rate for carpenters is $3.00 per $100.

Many sub-contractors furnish certificates for Workmen's Compensation Insurance that are not correct, will not apply, and in the case of an accident the builder-owner would be liable. Quite a few subs, in my years of experience, have answered the question "Do you have Compensation Insurance?" by stating "Yes, with such-and-such a company." What they mean is that they have hospitalization, life, car liability, insurance on their home, but they do not have Workman Compensation Insurance. Even insurance certificates from insurance companies sent directly to you must cover the men working on the job. Sometimes a mistake is made and the papers are for other trades than those working. Possibly, the certificate is outdated, not paid for, or any other number of sins make it worthless.

The answer to all this for a one-house subcontractor is to get the workman's compensation certificate for each subcontractor sent to you from his insurance company before he starts work. Then anyone who hasn't compensation insurance will have to be covered by a special policy, issued for just such a situation. A home owner had this done, at a cost of $4, for coverage on a friend. He was a retired carpenter who spent an afternoon cutting off the bottoms of his inside doors so that they would clear newly laid carpeting. If this carpenter had injured himself, no matter what the circumstances or whose fault, the owner would be liable for the man's injury, without the insurance. This is why one needs to have workman's compensation insurance when building for oneself.

However, workmen's compensation will not cover the builder-owner or children or friends on an "inspection tour" or while they are doing work on the home, so a third policy, a liability policy, to cover outsiders (non-workmen) and himself is needed by the home builder.

Further, let me point out, one insurance agent should handle the total program. The reason for this is that if two insurance companies get involved, one for the builder's risk policy and one for the home owner's policy, it is difficult to decide which company to bill in event of a claim and sometimes neither company will accept liability.

An example of this is a young couple who lived in a house for six months and noticed the large bathroom mirror was chipped. The home owner's policy man said it must have been the builder's fault because how could a mirror be chipped in normal living? He contended it was faulty installation that chipped it. The builder said it was not chipped when the people moved in. So everyone denied liability, a situation which could have been avoided if only one insurance company had been involved, with both policies. We solve this problem when building for people by encouraging them to use the same company that wrote our builder's risk insurance, for their home owner's policy.

33

13

PERMIT APPLICATION

To obtain a permit the first thing a professional builder does is call the local building department and inquire as to what is required to build a home. The further out in the country one builds, the less complicated is the procedure. In some areas, the process is so simple that all that is necessary is letting the building department know where the house is to be erected so that the improvement can be put on the tax roll.

The more complicated systems require the licensed subs to take out their permits first, before the general contractor can apply for his permit. These sub permits include plumbing, heating and electrical.

Well-septic plans must be approved, and sewer and water hook-on or taps paid for. Once all this is completed the permit for residential construction will be issued to the builder, or builder-owner. Quite often, zoning is involved and with all these items the cost in time can be two months of waiting for many details to be approved. The words, "Back to the drawing board," are used; and it is time-consuming. The cost of permits depends on the tax structure of the community. A well-to-do area which gets most of its tax money from a steel mill may charge $10 for all permits; another area next to it may require a total of $210. People who live in the area feel that they have developed all the other improvements — schools, city hall, and library, etc., so the $210 for permission to build is justified. Sometimes the community is short of funds and see high permits as a means to finance something without voting approval. On a market sales value, the $210 permit area building will be that much cheaper than the lot across the street in the $1,000 permit area, all other factors being similar.

Many areas require payment of a sewer or water tap fee before building even though there is no sewer or water available, and septic and well must be installed. This is done so that when sewer and water mains are put into the area money is in escrow to pay for hooking up. Other fees involve hauling permits, bonds, escrow deposits, curb cutting, and tree planting.

Watch for these variations before purchasing the lot and be aware that the lot is not necessarily a bargain because the permits are high. Thus, a five-acre parcel may be $2,100 cheaper on one side of the street than on the other side, the difference being in the permits. For those who are not familiar with the permit business, it is difficult to realize that a $1,200 sewer tap permit is just for permission to tap the sewer. It does not include the labor and material used in tapping the sewer, and bringing it to the house. Yet fees of that size are not uncommon.

14

SEPTIC SYSTEM

Today's changes are so rapid that a business man once told me, "How you run your business in the morning may not be the same in the afternoon." This is true of building, for overnight the shift from subdivisions with sewer and water to houses on acreage occurred. This means that wells and septic tanks must be added. A person involved in building or subcontracting his own home will be faced with a private septic tank and well system in 90 percent of the cases. The requirements for these change constantly.

Some communities require extensive engineering, soil borings and a detailed presentation of the sewerage disposal system. This is all overlaid on the topo, and done once by a survey engineering team, then field inspected by the sanitation engineer in the community.

Costs are relatively standard for a sewerage disposal system design, one engineering company charging about the same as the rest. However, if soil borings have to be done, get an estimate first as costs can be high, and you might not realize what prices are.

The well location must be pinpointed on the topo, especially noting the type of pump, and the distance between well and the hub of the septic tank, called the differential. It is best to call the building department or state building department before starting and ask them what the differential is. As an example, 50 to 200 feet are the figures quoted in our state, but the different counties have the responsiblility of increasing it if they feel the soil conditions warrant a greater distance.

To clarify this, let us take an example. The county says the differential is 150 feet and a 4 inch drilled well is required with a sumbersible pump, a 1,000 gallon tank and 300 feet of tile. No top soil boring or other requirements are involved — just the above specifications with which to comply, on final inspection. The big problem lies in the property frontage, which is 100 feet. The house on one side has a well located in the front adjacent corner and a septic system in the rear adjacent corner, while the house on the other side has the reverse situation. The 150 feet differential applies to the well and septic on the property next door, too. Thus it is impossible to build on this property. The owner of the land will find out the hard way that all he can do is put it up for sale again at a bargain price or hold it until technological improvements erase the 150 foot differential. A peice of property like this often is sold once a year because it is a bargain, available, and buyers do not know the requirements of the community when they buy it at a bargain.

15

WELL

Well estimates are also a must. They should describe in detail the pump size, casing costs per size, installation cost of line from well head to inside the house and per foot drilling charge. Final estimates usually depend on the depth of the well. This can, and often does, vary almost 100 ft. from the estimate. A good way to check the estimate is to ask people in the area what their well depth is. Local health departments sometimes have water table maps and can be helpful here.

Modern water systems specified by the health departments require a submersible pump. That is, a pump which is let down into the well. The water goes just one way, up and out.

Jet pumps and circulating surface pumps have the disadvantage of possible contamination of the water because they force water down to bring up more water. It is this two-way action that is risky from a health point of view. Also, because they are in the house, jet pumps are noisier than submersible pumps.

Where good water is close to the surface, simple drive points will work and are very easy to install and connect up to a jet pump yourself. Check to see if you can install your own well system.

The well driller should be chosen for his ability to service and maintain your system as well as his installation cost price.

Our well-driller has four drilling rigs, a 220 volt generator, air compressor with jack hammer, and all trucks radio controlled. One January day, a customer called me to report his well-head frozen. The driller was able to go right out, cut through three feet of frost and repair the break with his equipment. A well-man less equipped would have to subcontract to another driller or rent the equipment to do the job, and one tradesman does not have much enthusiasm fixing another one's problems.

LOT MAY NOT USE WELL OR SEPTIC BECAUSE OF A 150' SEPARATION RULE

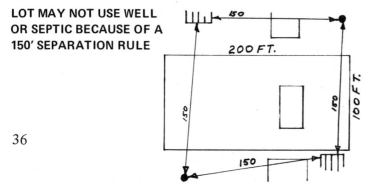

NO LOCATION FOR SEPTIC OR WELL ON THIS LOT WITHOUT INTERFEARING WITH THE WELL OR SEPTIC ON THE LOTS NEXT TO IT

36

Drilling and Pump Contractors

PRICE PATTERN ON 4" WELL

"All prices subject to change without notice"

per lineal foot black pipe. **

per lineal foot galvanized pipe. **

for a tempered drive shoe.

This includes pipe, labor, setting of strainer, development, and dynamite.

WELL STRAINER will be used and they are made with either RED BRASS or EVERDUR metals.

4 ft. Everdur 3" I.D. Strainer with fittings
cost—   per foot.

4 ft. Red Brass 3" I.D. Strainer with fittings
cost—   per foot.

4 ft. Stainless 3" I.D. Strainer with fittings
cost   per foot.

Strainers may vary in length because of sand and gravel sizes. In fine sand we recommend 5 to 8 ft. of strainer to insure a good producing well. Some gravel may not need a strainer or produce an adequate amount of water through 3 ft. of strainer.

Sales Tax will be charged on all material used.

On our 4" wells you receive a WARRANTY in writing for five years on material and workmanship. If strainer plugs we change it free of charge. We guarantee water but not the amount or potable. In other words pin a driller down to how much and what kind of water. If he does not commit himself he is not guaranteeing you one thing.

**Pipe specification A.S.T.M. Schedule 40—new   1200 P.S.I. test. Reamed and drifted using heavy reamed and drifted couplings—weight 11 lb. per foot. Check brochures for more claims on our pipe.

We maintain four complete drilling rigs and tools plus a truck with hydraulic power hoist for pump service calls. We have a 1250 gallon tank truck with a 90 G.P.M. pump for developing your well, and a TRACTOR BACKHOE for any needed digging and trenching on pump installations.

"ACROSS THE ROAD AND 100 FT. FARTHER DOWN?"

When your well is twice as deep as your neighbor's what are you to say?

"You drilled right down through the good water!"

Or, "You're just out to stick me with a big bill!"

Or, "You * @!*¼-!&****

At times like these, a good, sound explanation sure helps smooth things over. An angry customer can break our reputation because he doesn't know the simple facts. Don't hurt our future sales. This is what you need to know.

GLACIAL DEPOSITS

We've all heard of the "ICE AGE", and even though the age itself in North America is 10,000 years gone it can still give a driller trouble. When the great continental glaciers covered four-fifths of North America (and Europe, as well) they literally tore the surface off the northern countries and then dropped it farther south.

We run into this situation in the northern half of the United States. As we drill down through clay and silt, you'll hit good water in a pocket of sand that an uptown geologist will call a "lens". (See Figure 1.)

The lens might measure a thousand yards or a hundred feet across. There's no way of knowing without drilling down to see. In one place we will get enough water for a single residence or a farm — but twenty feet away the lens might have petered out.

Fifty feet away we might go all the way down to bedrock and get nothing. Or we might hit a long buried alluvial deposit with enough water to float the Queen Mary.

ALLUVIAL DEPOSITS

Ever since the earth was a pup, its surface has been constantly changing. These days, we can go down just about anywhere, and find what once was a river, lake or ocean.

The sandy beaches of bygone lakes and oceans and the coarser deposits of old river beds, are where you'll find water. You can have miles of leeway in the broad sand and sandstone deposits left by larger bodies of water. The buried coastlines of long ago oceans are just as big or bigger than the ones you see today.

But ancient buried rivers are a different story. If we miss the bed you can end up with a dry hole in silt or bedrock. (See Figure 2).

Experience, observation and good old-fashioned horse sense give us a working knowledge of the area. If there's a narrow band of good wells with nothing on either side, it's entirely possible we are dealing with an ancient river bed. We let a customer know what his chances are beforehand.

In the western United States alluvial fans are a major water source. This happens because the mountains were worn away. (See Figure 3).

ROCK FORMATIONS

Even in rocky areas it is possible to find a good alluvial deposit of

sandstone that yields plenty of water. Then one day we will be drilling where the sandstone still ought to be and we will find it way up higher than it was the last time. Or we will go down, down and down and we won't find it at all!

When that happens we can make a safe bet that there has been an earthquake in the area at one time. Pressure has made solid rock shear off so that one side went up while the other side went down. (See Figure 4).

And that's just a simple "fault" structure"! You may even run across some complicated "folds", especially if you're drilling in the eastern mountains, but don't let it worry you.

We let our customer know what we are getting into and what his chances are. With every well we watch what we are doing and keep a record.

We know the more we tell our customer about his well the more respect he will have for us.

A driller that knows how to screen a well can either drill in glacial deposits or bedrock areas, but too often a rock driller does not even have tools for setting a screen in a well. Be sure you get a sand and gravel driller because wells that are made in glacial deposits 99% need screens. If your bedrock is a poor producer of water how your odds go sky high on getting a deeper well or salt water with a rock driller.

A hole in the ground is cheap, but a good well is worth money.

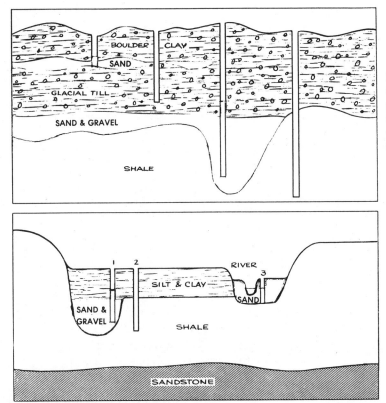

Glacial Deposits

FIGURE 1. Cross section of a glacial deposit resting upon water-deposited sand and gravel over bedrock. This diagram shows how two good water wells and two dry holes may result depending upon location of wells.

Ancient Rivers

FIGURE 2. Cross section of a river valley. Well 1 goes through an old flood plain into a still older river bed. Well 2 misses the river bed and is dry. Well 3 goes into the presently existing flood plain.

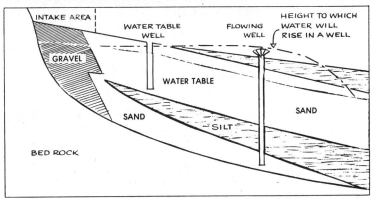

Alluvial Fans

FIGURE 3. Typical cross section of an alluvial fan. This fan is built in two steps. Drilling into the lower sand body results in an artesian well because the water pressure is kept up by the layer of silt confining it.

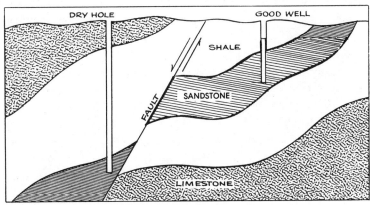

Rock Faults

FIGURE 4. Geologic cross section through water-deposited rock formations that have been folded and faulted. Arrows show direction of movement on either side of fault. Note how down-thrust side "dams" water in sandstone layer.

16

ZONING

Zoning is a highly controversial type of rule. It is quite new which means that one could build just about whatever he wanted anywhere he wanted in rural areas up until the last few years. Even some large cities have just recently reviewed and revised zoning requirements. Today checking zoning is the responsibility of the building department. The plans and specifications must be approved by them first before any further processing can take place.

Often a subdivision committee performs this inspection before the building department does. In other words one must present his display (drawing and application) before this local subdivision committee and ask for their approval before requesting a building permit. Deed restrictions are overlapped by the subdivision's, then by the municipality, so that no matter what the deed says the city has the last word, based on the subdivision committee recommendation. This committee is made up of three or four members who live in the subdivision.

Again these rules or restrictions vary all the way from the community simply wanting to make sure the people build a house to live in, and not a barn without sewer or water into which they move (it has happened). The other extremes are a combination of architectural harmony, color scheme and anything else someone can dream up that would be a possible reason to say "No, you can not build," especially if politics are involved. Or the theme of the subdivision may be Early American which is the only type of construction they will approve, and it does make a nice looking subdivision if everyone follows the style with variations.

All zoning today is restrictive and takes one direction. That is, toward more costly construction. As time goes on, the people who are already living in the area get together and decide that the next people building around them will have to put up larger and more expensive homes. Thus a young couple will buy a lot that has a deed restriction requiring 800 square feet of floor space. This is a saleable feature and the seller is careful to point it out. However, two years later, when the people start building, they find that the subdivision committee has decided that the minimum size will be 1,500 square feet and the municipality has over-ridden this with a 1,700 square foot size on all 100-foot front lots, which includes their lot. Usually the subdivision is more restrictive than the government body.

40

Now, how can they do this? The legal reasons are beyond the scope of this book, but some of the rules and requirements of zoning are so demanding and absurd that they require challenging. When this happens one goes before a zoning committee or board and presents the case, asking for a variance of the zoning rules based on the inability to comply, or hardship. All these requirements are very vague and depend on the whims and disposition of the people on the board of zoning appeal.

If the appeal is turned down, the next step is a civil suit which can be dragged out for as long as 10 years. By that time everything has changed and, win or not, you may no longer be interested in building because your needs have changed.

A land use permit is another type of overlapping building or zoning permit, issued on a township or small city level. This means that the local small governing agency wants control over the building that is proposed in it's area, which is larger than a subdivision. Thus, a land use permit must be applied for and issued by the usually part-time local persons in authority who lives in the area. Often these are groups of five people with at least three present for a 'yea' vote. This can be a big job, as it is customary for these people to meet at their convenience and if they are deer hunting or vacationing, the permit must wait. The county or next government unit that issues the actual building permit will not do so unless this preliminary step is completed.

Information for this permit varies, but is mostly a question of identifying who you are by name and address, where you are from, and what type of building you are interested in constructing on your land.

Zoning does have a good side in that, in general, it does assure somekind of conformity in having houses approximately all the same size on the lot, about the same distance from the street, and in relatively the same cost bracket.

It has abuses and they are often politically inspired, changing from residential to commercial, to industrial, property over which the actual owner has little to say. Or, by increasing the size of the house from 1,000 square feet to 1,500 square feet, this eliminates the dream home of the young lot owner who has saved all his life and can just afford to build a 1,000 square foot house. This is an excellent reason why the permits should be taken out before the lot is purchased, if possible.

In one case recently, a farmer wanted to build on his 30 acres. The community zoned it industrial, and it took a year plus a new political regime, to get zoning back to residential property so that he could build his home.

These rules are not complicated or difficult, it is just that they are new and people do not know about them. Half the existing houses in an area are shacks or modified garage homes. Then someone decides to build and is shocked to discover that only a large modern contemporary style house with two-car attached garage will be accepted.

If you get into a positon where you will want a variance certain steps may be taken to increase your chances of success:

1. If the present owner is well known in the community and well established, get him to file for the variance and then you or your attorney represent him.

2. Be prepared
   a. Dress neatly.
   b. If you are going to live in the home, impress that on the members.
   c. If sketches are needed, make sure they are neatly drawn and measurements are reasonably accurate.
   d. Get a picture of a like building incorporating the variance. The board members usually cannot visualize blueprints — they often are not trained people.
   e. If possible and time allows, get the neighbors to sign they are in favor.
   f. Familiarize yourself with the area and list any nearby similar variance as the one you want.
   g. Do not treat the board as an antagonist. Be friendly.

17

BUILDING PERMIT

Part of the cost of building permits, or all of it, is used to provide inspections. The chart here illustrates just how far a community can go in requiring inspections. This is one community's actual inspection schedule:

BUILDING
Footing Open
Rods in Footing
Grade
Ready for Backfill
Floor Joist
Before Subflooring is Applied
Truss on Ground
Rough Framing
Lath
Final Grade
Final

PLUMBING
Septic Tank & Field
Sewer
Crock to Iron (Where sewer enters house)
Inside Drain
Shower Pan
Rough Plumbing
Water Test
Well Water Test (In case of well)
Final

ELECTRICAL
Rough
Finish

CONCRETE
Rail
Sand
Rods or Wire
Basement Floor
Sand
Rail Outside (Walks and Drive)
Final

HEATING
Rough Heat
Insulation
Final

At the other extreme is individual subdivision registration, which simply is the obtaining of a permit with no inspections at all, and no code requirements. A tax appraiser or accessor checks the building and puts it on the tax roll. That is all there is to it. In building with this simplified system, it would be best to follow another community code because eventually, when the governing unit is better organized, it will want buildings brought up to date, and building to a code does give a high safety factor and is cheaper in the long run. Sometimes a state code applies to a community that doesn't have one.

When the permit is issued, the city clerk collects the money and gives the builder receipts for the various permits and a weather sheet. This is a cardboard sign to be posted on the job and says "Permit to Build" across the top. It lists inspections required below and a place for the various inspectors to sign and put a date. Also printed on the weather sheet are the office hours when to call for inspections, and where the building department is located. If it is not printed on the sheet, make a copy to help you in calling in.

Some building departments run a tight, well-organized, pleasant type of operation. They record your requests for an inspection, give you a call or code number over the phone as a receipt for your call and if you have special questions they will let you talk to the inspector and make arrangements to meet him on the job by appointment. They also have radio contact with inspectors in the field and can arrange to have an inspector check a concrete pouring job within 15 minutes rather than have the men go home from the railing job early, and wait an extra day for the open trench inspection, before pouring the concrete. If an inspection passes they record it in their book and leave a waterproof dated sticker on the building coded green with their signature, date and the type of inspection marked OK to proceed with the work. When it is not in order a red sticker, dated and noted as to what is wrong, is left if the violation is not too complicated. On something more difficult the inspector requests that the builder call him. He does everything he can, within reason, to help the builder.

Now comes the other type of building department. Politically appointed inspectors who have authority and no experience can really give you a rough time. Your house is costing you $10 or $12 a day in interest money and construction funds and they can hold it up by simply refusing to give inspections. There is no record of your requests for inspection, or the inspector does not show up. One small town inspector who insists on inspections does not answer the phone and has the treasurer take over when he is on vacation (about every two weeks). Then whoever finally makes the inspection refuses to sign anything, saying it is okay on the phone and that he will mark it down. Two months later, no one knows what you are talking about, but they want, for instance, the off-street parking, which had passed verbally, moved some place else; and they want the wall-board cut out to expose the plumbing stack that was never checked, according to their records. Your records may show that they said it was all right to go ahead with construction and that the rough plumbing was in order. But the plumbing inspector is a part-time man and only works after the evening dinner hour, so he couldn't find the weather sheet at the time to sign it. How can you prove he checked it if their records show he did not? So, you just cut the wall board out and expose the plumbing stacks. You can't fight city hall.

This brings up another point. People say get a lawyer, make them do this or that. No matter what you do to the politician in power, people will stick together, supporting each other. They have a municipal lawyer to consult; you have to pay yours. They have

44

BUILDING PERMIT

DEPARTMENT OF BUILDINGS AND SAFETY ENGINEERING,

PERMIT No.

City

of

THE BUREAU OF LICENSES AND PERMITS HEREBY GRANTS PERMISSION TO

DATE _____ 19____

NAME _____ ADDRESS _____

CONTRACTOR'S STATE LICENSE

TO _____ ( ) STORY

NUMBER OF APARTMENTS

ON THE _____ SIDE OF (BUILDING NO.) _____

ST. OR AVE.   ZONING DISTRICT

BETWEEN _____ AND _____

ST. OR AVE.

LOT NO. AND SUBDIVISION _____

SIZE

BUILDING IS TO BE _____ FT. WIDE BY _____ FT. LONG BY _____ FT. IN HEIGHT AND SHALL CONFORM IN CONSTRUCTION

TO TYPE _____ USE GROUP _____ BASEMENT WALLS OR FOUNDATION _____

REMARKS:

APPROVED BY _____

®

CUBIC FEET. _____ ESTIMATED COST $ _____ REGULAR FEE $ _____

ZONING FEE $ _____

OWNER _____

BUREAU OF LICENSES AND PERMITS

ADDRESS _____

By _____

☞ THIS PERMIT CONVEYS NO RIGHT TO OCCUPY ANY STREET, ALLEY OR SIDEWALK OR ANY PART THEREOF, EITHER TEMPORARILY OR PERMANENTLY. ENCROACHMENTS ON PUBLIC PROPERTY, NOT SPECIFICALLY PERMITTED UNDER THE BUILDING CODE, MUST BE APPROVED BY THE COMMON COUNCIL. STREET OR ALLEY GRADES AS WELL AS DEPTH AND LOCATION OF PUBLIC SEWERS MAY BE OBTAINED FROM THE DEPARTMENT OF PUBLIC WORKS - CITY ENGINEERS OFFICE. THE ISSUANCE OF THIS PERMIT DOES NOT RELEASE THE APPLICANT FROM THE CONDITIONS OF ANY APPLICABLE SUBDIVISION RESTRICTIONS.

MINIMUM OF THREE CALL INSPECTIONS REQUIRED FOR ALL CONSTRUCTION WORK:

1. FOUNDATIONS OR FOOTINGS.
2. PRIOR TO COVERING STRUCTURAL MEMBERS (READY TO LATH).
3. FINAL INSPECTION FOR COMPLIANCE PRIOR TO OBTAINING CERTIFICATE OF OCCUPANCY.

APPROVED PLANS MUST BE RETAINED ON JOB AND THIS CARD KEPT POSTED UNTIL FINAL INSPECTION HAS BEEN MADE. WHERE A CERTIFICATE OF OCCUPANCY IS REQUIRED, SUCH BUILDING SHALL NOT BE OCCUPIED UNTIL FINAL INSPECTION HAS BEEN MADE AND CERTIFICATE OBTAINED.

POST THIS CARD

SEPARATE PERMITS REQUIRED FOR ELECTRICAL AND PLUMBING INSTALLATIONS

BUILDING INSPECTION APPROVALS	PLUMBING INSPECTION APPROVALS	ELECTRICAL INSPECTION APPROVALS
1 DRAIN TILE AND FOUNDATION_____	**1** BUILDING SEWER (A) SANITARY_____	**1** ROUGHING IN_____
DATE_____	DATE_____INSPECTOR_____	DATE_____
INSPECTOR_____	(B) STORM_____	INSPECTOR_____
2 SUPERSTRUCTURE_____ (PRIOR TO LATH AND PLASTER)	DATE_____INSPECTOR_____	**2** FINAL INSPECTION_____
DATE_____	**2** CROCK TO IRON_____	DATE_____
INSPECTOR_____	DATE_____INSPECTOR_____	INSPECTOR_____
3 FINAL INSPECTION_____	**3** UNDERGROUND STORM DRAINS_____	SAFETY ENGINEERING APPROVAL
DATE_____	DATE_____INSPECTOR_____	APPROVED_____
INSPECTOR_____	**4** ROUGH PLUMBING_____	DATE_____
	DATE_____INSPECTOR_____	INSPECTOR_____
WORK SHALL NOT PROCEED UNTIL EACH BUREAU HAS APPROVED THE VARIOUS STAGES OF CONSTRUCTION.	**5** WATER PIPING FINAL INSPECTION_____ DATE_____INSPECTOR_____	INSPECTIONS INDICATED ON THIS CARD CAN BE ARRANGED FOR BY TELEPHONE OR WRITTEN NOTIFICATION.

everything working for them, when you try to prove something. Delaying techniques and court procedure can be strung out to such an extent that after two years, if you win, they can say, for example, "Now you can move in and you do not have to change the driveway." But you have to pay your lawyer and interest on the money tied up in the house for two years. And if you lose you are a double loser because of the legal fees and the time element.

The final certificate of occupancy is issued upon completion of all inspections. Here again, some communities have penalty clauses that read $5 a day fine if a house is occupied without a Certificate of Occupancy, or even more in some cases. Others have no requirements for occupancy except that the roof be on. This helped me once in that excessive thieving and vandalism were occuring on one house. By moving the young people in and building the house around them I was able to complete it without further loss. It was summer and the people saved a couple of months' rent while making it possible to complete the house in record time because I did not have to go back for replacements and repairs on theft and vandalism.

The plan I like best involves a bond held by the community. When the house is completed except for one or two items, the bond will cover a guarantee that these will be taken care of and the people can move in. Then, when the work is done, an inspection is made and the bond is returned. Of course, everything to be completed has to be written down or there is a tendency to add items after the house is occupied.

The Certificate of Occupancy defines the legal line between a house under construction and one ready for occupancy. All C.O. are subject to being revoked for any changes in plans or noncompliance with the building codes that an inspector overlooked; or any fraud used in obtaining one. Some communities are issuing certificates that expire every two years, so that a renewal must be obtained every two years through a new physical inspection of the property.

It is very important to have a permit, even a temporary one, if you move in, because a home owner's insurance policy is based on the property being approved for occupancy. It would be void if you lived in the house and did not have a special rider or some clause to take the place of an official Certificate of Occupancy.

There is a new trend that requires plumbing, heating and electrical permits to be taken out by licensed, registered master mechanics only, and all inspections must be requested by these people directly from the building department. The latter will not take the requests from the builder or owner for these inspections or permits.

The time to get all of these permits and the okay to start a house varies from one day to slightly over a month.

When the work is completed, call the various tradesmen to have them get their final inspections, because this often takes time, and waiting till the last minute will really be taking a chance in case some adjustments must be made. One violation will hold up the whole permit for occupancy until it is corrected and reinspected, a procedure which should be done right away as the inspector can come up with some surprises.

One job done with hot water heat had a small boiler jammed in the closet when ready for final inspection. A week before, the city had passed a ruling that all boilers should have a

3-foot clearance all around. This necessitated moving the boiler to the garage and building a room around it. Fortunately, the heating and plumbing inspections were requested as soon as the work was completed so that the alterations were made immediately while the rest of the construction went on. Thus, the whole job was completed as per contract.

The more you can get the building inspector to hang around your job or look it over, the better off you will be, because any suggestion he makes usually indicates how he wants it done and he is the boss when it comes to this.

IMPORTANT NOTICE !

This Building cannot be legally occupied until the building owner has a certificate of compliance issued by the Building Inspector.

DEPT. OF BUILDING & SAFETY
Phone 1928   **Howell, Mich.**

Final inspections, both municipal and mortgage certification should be allowed even with the elimination of stealable items. Some of these are:

Storage shed	Appliances
Disposals	Storms & screens
Range Hoods	Carpeting

In other words, to prevent theiving the community and mortgage company should allow you to move in, then install these items.

Home owner building permits are available to any lot owner, who indicates he will live in the house, but many mortgage companies today will not lend to do-it-yourself builders because of the high rate of noncompliance they have experienced previously. Rarely a lender will qualify a borrower by requesting he take courses in building at the local high school or 2 year college level, almost the equalivent of qualifying for a builder's license.

The main problem for building departments comes when, do-it-yourselfers run out of money and looks for someone to blame.

18

EXCAVATOR

The excavator is the man who goes in first. He has a plot plan and can follow it to stake out the excavation. You as the builder should work with him in checking the layout before he digs. Also, check the grade; this is the distance the house goes above the ground from some check point such as the height of the crown at the center of the road. In other words the house grade is described as two feet above the center of the road, with a hundred foot setback. The center of the road acts as the main check point and will not vary more than four inches for the life of the house. Sometimes four inches of asphalt is added to the road height over a period of years.

If there is a sewer, the digger starts there. Tapping the sewer requires an inspection, and reservations must be made a day ahead of time for the inspector to come out, say, between 8 and 10 in the morning, or whatever time the sewer will be open. When the sewer is tapped the excavator can go ahead and dig the trench for the sewer line. If water is available, he can make a shelf next to the sewer to hold the water line too, and be careful to get inspection on this, sometimes by another inspector. Starting at the tap, he lays the sewer line up to the basement footer line.

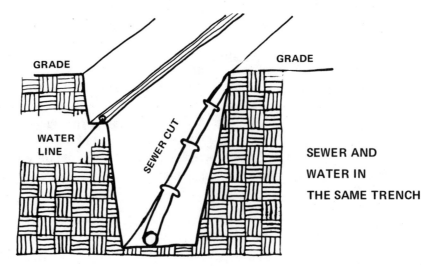

GRADE

GRADE

WATER
LINE

SEWER CUT

SEWER AND

WATER IN

THE SAME TRENCH

Keep the sewer line a foot and a half below the basement footer level, as the sewer line goes under the wall. In digging the basement a six-inch crown can be placed on the floor to help save on filling in with sand later on. However, if the ground is clay it doesn't pay to do this as the extra effort required to dig the drains through heavy clay costs more than the saving on sand. The excavation is dug two or three feet larger all around, depending on how well the ground holds up. Dirt from the excavation must be piled so that two opposite corners are open to transit trucks.

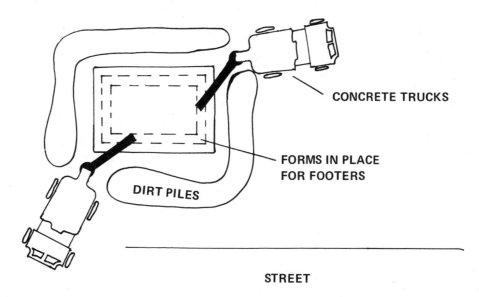

CONCRETE TRUCKS

FORMS IN PLACE
FOR FOOTERS

DIRT PILES

STREET

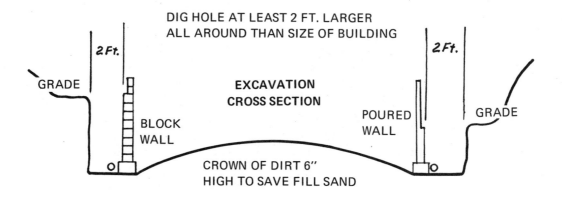

DIG HOLE AT LEAST 2 FT. LARGER
ALL AROUND THAN SIZE OF BUILDING

2 Ft.

2 Ft.

GRADE

EXCAVATION
CROSS SECTION

BLOCK
WALL

POURED
WALL

GRADE

CROWN OF DIRT 6"
HIGH TO SAVE FILL SAND

After the sewer is inspected, a permit or receipt must be signed by the inspector to certify the inspection. Then the excavator backfills the sewer up to the last crock just before the footer line. A 2x4 or other large stake placed here will indicate the end of the line so that the plumber or footer man can easily locate it. Often the excavator includes in his contract the backfilling after the basement walls are up.

If a home is built without a basement, a crawl space is excavated the same way as a basement, but not as deep. In effect it is a miniature basement. However, foundations may simply be a trench around the perimeter. The depth of the footer below ground level or grade depends on the local code. This digging can be done by hand or with a trencher. Costs are about the same — the trencher is fast and expensive, hand labor is slow and expensive. If roots or large stones are frequently encountered, then hand digging may be the only alternate.

Basement or crawl space require drains under the house. These are called inside drains and are installed by the excavator or plumber. Thus the excavator will do labor and material on the basement, sewer line, water line, inside drains, stake out the basement, backfill the sewer, and backfill the basement. This should all be written down and signed as work agreed upon so that their will be no misunderstanding.

There are other special things that the excavator can do. He can pump water, straw down basement floors to keep them from freezing, set sewer traps, remove stumps, cut and replace paving, and take out old footers that are in the way. All of these should be agreed upon ahead of time so that everyone understands just what is not going to be done, what is going to be done, and what it will cost. Again, this is done because of the high cost of labor and the risk of some jobs for the excavator. Do not hesitate to ask how much to take out a tree, or is it part of the job, or will it be quoted separately as so much per hour, who will get and be responsible for the inspections? Many builders save money by doing as much of the calling for inspections and supervision as possible, paying only for the actual shovel time.

If the ground is very hard, a sump pump hole can be dug with the back hoe in the best location possible to keep noise from the pump at a minimum and stilll have the piping from the house to the drainage ditch as short as possible. The sump pump should be located under the kitchen, the living room, or the porch, not under bedrooms where the on-and-off noise might disturb sleepers.

Plumbers set traps and do inside drains in a great many areas. It is part of their job because they are responsible for the final working of the plumbing. However, this is a 10 minute job and some excavators and wall footer men are good at it. They do it rather than involve the plumber with a lot of travel time for a little work. If it is described as plumber's work, the builder leaves it up to him to pay for, or arrange a deal with the excavator. Decide ahead of time who will be responsible for the trap setting and inspection if necessary.

A caution about tree removal, never cut one off even with the ground. To remove the roots and allow a bulldozer or back hoe leverage, leave the tree trunk at least 4 feet long above the ground, thus with a minimum of digging around the base of the roots the machine can use the trunk sticking up as a lever and push-pull, loosen the roots enough to extract the stubby trunk and roots easily. This is very simple compared to digging the whole thing out, a necessity if it is cut off flush with the ground, or too low for the machines to obtain a purchase.

19

FOOTER AND BASEMENT WALL

Two types of foundation walls are used in housing construction, block or solid poured concrete. Usually the footer base for the wall is put in by either of these subcontractors. However, there are breakdowns or divisions of labor where the footer is put in by one subcontractor, and the wall by another. Whoever it is, should be sure, when there is a sewer, that the plumber's trap is set below the footer into the house area, and inspected if necessary, especially the joint between the crock and iron. Various communities require the plumber to do this and occasionally the union has a requirement stating that only the plumber can do sewer work. The same is true of inside drains. In some areas plumbers put them in; in others, the excavator or a footer man with labor does the work. From one community to another these things can vary, so that on one side of the street it is done one way and on the other side differently.

The footer for a poured wall should have a "V" in the center top made by putting a 2x4 on edge in the wet cement. This locks the poured wall in place. All footers must be on solid ground, with all loose earth out of the forms before pouring. If there is any question about the virginity of the ground — whether the dirt has been disturbed, use reinforcing rods or a spread footer, to make sure the basic foundation will not settle. When ground is not virgin soil, it has been filled in, or it may have soft peat which will not hold a regular footer, if this is the case, the local building code or inspector will help advise as to how to handle it, usually by digging down to solid ground. However, piling can be used, or wide steel-reinforced floater footers, depending on the soil compaction. These remedies for nonsupporting soil are complicated and require engineering designs, or the special attention of the building department. Sometimes the building department is well-enough acquainted with the problem that they can point out what type of foundation is needed and specify it.

There is an inspection called a rail inspection, or an open trench inspection in some areas. It occurs before the concrete is poured. The inspector should be contacted ahead of time so that the footer is railed, inspected and poured; all in one morning. In our area, the inspectors will accept two or three inspection holes dug down next to and under the footer, so that they can see how deep the concrete is, and look under it. This hole inspection system is used only when open rail is not timed just right and the inspector doesn't want to hold up the job.

A poured basement is made by using a series of steel, aluminum or plywood forms 8 or 10 inches apart, filled with concrete. The form is stripped when the concrete has set. All openings must be decided upon before the wall is poured, then wood boxes, tin cans, or thimbles put in between the forms where these openings are to be in the wall. The holes to consider are:

A heat thimble in the chimney is a hole for the furnace pipe to enter the chimney. Directly under this is a smaller thimble for a hot water tank vent. Then, a 6 inch thimble for an incinerator opening into the chimney. Many communities are requiring that residences have garbage disposers and incinerators to combat the high cost of collecting garbage and trash. If not used, this incinerator hole can be mortared over for future use, requiring only a tap with a hammer to open it up.

A Sump Pump discharge hole should be located three inches below grade and over the sump pump location. It can be a 3 inch sleeve or thimble.

A well hole, if required, is located 18 inches above the basement floor, deep enough outside so that it won't freeze. Put it in a place where the pump or tank noise will not disturb anyone. It should be 4 inches in diameter.

The septic hole is described by the plumber. He will tell the builder just where he wants it and what size it should be and this information then is conveyed to the wall pour man.

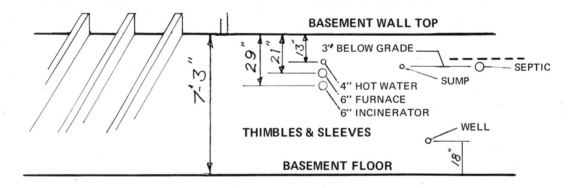

Windows and doors require wood boxes the same size nailed in place between the forms.

Porches and garages will need footers with blocks to bring the porch and garage foundations up to grade away from the loose filled earth.

The block wall doesn't need preplanned holes, except chimney ones, because the different trades can cut their holes through the block to suit themselves with relative ease. Some builders do this with poured basements too, but they have access to an air hammer or electric hammer which makes the job simple. It doesn't make sense to cut through a poured basement wall with a hand hammer and chisel because of the cost.

The block layer will need water and the materials for the wall construction. He will furnish all this and give a completed price for the whole foundation. It is a good idea to help him with the water, because he may not be familiar with the neighborhood, and it could be a problem. Areas that are all gravel or sand where water would not be able to collect do not require weeper tile; but be sure to check with the local inspector. The block man can finish laying the blocks, then cement plaster the outside for waterproofing. Over this he puts tar or black mastic to complete the waterproofing. The basement should cure or set about a week depending on how cold it is. In cold weather, curing takes longer. This is true of block or poured concrete basements.

Brace the walls with two-by-eights, or special steel braces. After this is done, the tile is in place covered with pea gravel and the wall is waterproofed, the bank or local inspector will want to look at it. When the work has passed inspection, obtain a receipt or some kind of record for your file, and the basement is ready to backfill.

Backfilling requires an expert. It is so easy to hit the wall, punch it in or crack it. I would rather have a machine-operator do a half-way job and finish by hand, if the driver is not an expert. The bulldozer operator should stay 3 feet away from the wall at all times and never run along the length of the wall. He should roll the dirt into the hole by pushing it in at right angles to the wall. Any big lumps left by the machine should be smoothed out because many men will be working around the house on the outside, and it will help to have the dirt fairly even. Clean all excess dirt off the top of the wall, brick ledge or footer pads immediately. If it rains or freezes and the dirt sticks on these surfaces, it must be removed. Waiting just makes it harder to scrape off.

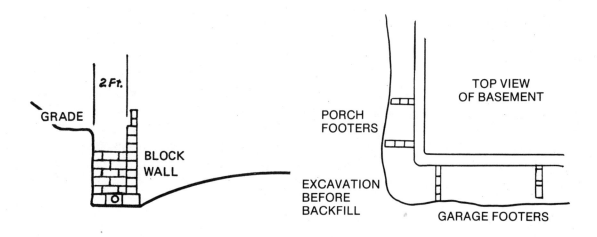

Put in blocks for porch or garage footers before backfilling. These are called lead walls.

20

ROUGH LUMBER AND OTHER MATERIAL

The rough carpenter is the next in after the basement or foundation is completed. Before he can start, lumber, windows, hardware, beams, stanchions, or anything special he might need must be ordered.

A lumber list can be obtained from the rough carpenter, which should be used in shopping for price. If a lumber list is obtained from some other source such as the lumber company, the man who furnishes the plans, or a specification company, have the carpenter layout man look it over and make sure it is adequate, then use this list without changes to shop prices. Lumber quality should be consistent with that used in the area. The local lumber companies will know what the area code requires; so will the rough carpenter subcontractor, which gives you a double check. When shopping, be sure the quality quoted on is the same on all the estimates, because there are great variations in grades of lumber.

It saves money to specify less expensive grades for backing, plates, doublers and places where the use doesn't really make that much difference. In these selections, a builder knows just what to do, while an amateur can be talked into buying lumber on the basis that it looks good, is stronger, costs more so therefore is better, or other emotional reasons that have nothing to do with the right lumber for the job. The carpenter foreman can also tell what grades to use.

Windows should be delivered on the job the day rough construction starts. This allows a positive check against error in that the rough openings for the windows can be checked right there by putting a window in place in the opening. If windows cannot be delivered on the day construction starts, the window company can make the rough openings on the blue print for the carpenter to follow, and the windows can be put in later. Sometimes, because of thievery, it works best the day the windows arrive to install them right away. Another system is to install the open aluminum or steel frames without glass, then put the glass in later. This cuts down breakage and theft, especially with constlier double seal or insulated glass inserts. Occasionally we take the windows to the job ourselves and install them immediately.

A word about theft: At this point, the building project becomes attractive for someone to walk off with lumber, or whatever else is not nailed down. Up to this stage, material such as cement block does not have a lot of use to the average person. Regardless, we immediately post signs on a beginning job which read: Call Police. Reward. Or, KEEP OUT, DOG, or ARMED GUARD. Such signs work wonders even when no dog or guard is present.

Labor on the job, skilled or unskilled sometimes is responsible for taking material. Trusses are quite difficult to haul off a building sight in a pickup or on a car roof once they are delivered. While 2 x 6 roof lumber will often disappear. Theft is a big problem and regular builders solve it by averaging the cost among all the homes built by them. THEY MUST DELIVER A COMPLETE HOME TO THE CUSTOMER. While a do-it-yourself house builder may not be able to absorb a big loss of stolen material. A local ad in the paper recently read, "Reward for 1000 lb. pre. fab. circular steel stairway stolen the night of June 11. Phone 274-8106." Scarey, but it shows that almost any thing can be taken.

Wooden windows, bow windows, bay windows and windows with fancy scallops and bric-a-brac on them can go in after the waterproof roof is on, to keep them from warping out of shape, or getting so wet it is difficult to prime, if rain comes.

Sheet plastic flashing should go over the window top if it is a two-story house, or the opening is on a gable exposed to the weather. Some codes also require this under the window, too. The reason for this is that a driving wind blown rain can follow various cracks and blow the water right through the wall, even up under the window. The plastic flashing prevents this. It is not necessary if the window or door is under an eave or overhang of any type.

The steel beam is ordered from the steel company and they will take the measurements right off the blue print. Stanchions to support the beam can be ordered at the same time. These are measured by stretching a line or string across the basement wall from beam pocket to beam pocket, then measuring from the stanchion pad to the string. This gives the height of the stanchion. Take an extra inch off when ordering the stanchions this length. As the stanchions are installed, steel shims between the bottom of the stanchions and the footer will make up the difference and provide a method of adjusting the beam for the exact height. There also are adjustable stanchions that may be used. A bearing plate one-fourth inch thick by a 8 inch square is used where the beam rests on the beam pocket in the cement block, or in the poured wall pocket. Shimming is done between the plate and the beam to bring the top flange flush with the wall.

Metal shields for termites are needed for some codes and these will have to be provided before wood is placed on the foundation.

Rough hardware includes nails, louvers, air vents, joist angles, braces, and joist hangers.

Electricity should be discussed, since the workman will have to have power to run the saws and electrical equipment. Most builders handle this with a portable, 3,500-watt generator that goes with every carpenter crew. When a drop line is used to borrow power from the nearest house, the voltage drag can be so great if the electric extension line is long, that blown fuses are common. To remedy this, the fuses are increased in size, but the voltage drop may still burn out the saw motor, or cause an overload in the house circuit. Some people have an electric or hydro box placed on a pole near the house, this is ideal but is expensive for one house. Once the house is up, the permanent line can be installed on a temporary permit. This provides power for all the following trades and even temporary furnace electricity for heat if needed.

21

ROUGH CARPENTER

The rough carpenter crew is made up of a layout man, saw man, cornice man, nailers and laborers. Team work is all important and as the rough lumber goes together every man knows his job and works smoothly with the crew, because they are organized. A house of 1,000 square feet can be finished in 100 hours of carpenter work, if the framing is not too complicated. Twelve men working eight hours and a fraction will finish the job in one day, or six men working two days. Usually the builder services the carpenters, watching them to see if they are short on any material. The faster the building goes up the less the servicing time, and the better for keeping weather damage and thieving to a minimum.

The big problem in theft is twofold: first, the value of the materials, and second, the interruption of the schedule. If someone steals 20 floor joist needed for floor support, all work must stop until they are replaced. If there is a wait for a day or two, the work may be held up by rain for a week, rain in the spring may not stop the actual work but will make the access to the job impossible because of mud and water. Sometimes a day is critical, making the difference between getting the roof on before a big rain or snow sets in, causing as much as a month's delay. Then too, any tie-up costs money in terms of the mortgage interest going on while the job is dormant. The name of the game is push and get the house up fast, or at least closed in.

A good crew can do a beautiful job of building everything on the deck. The deck is the subfloor, made usually of ½-inch plywood or ¾-inch boards. The side components of the building are built on the deck or floor, used as a giant workbench. By framing the house sides and ends, complete with gables, flat on the deck, then raising them up, much time is saved. The carpenters raise both sides up, and brace them; then frame and raise the ends and nail the corners together. Everything that can be done while the walls are on the deck flat is easy. Once the walls are raised up vertically, work on them requires scaffolding and ladders. Some crews can put windows in, cornice on, bays, porches, overhangs and even siding, when the front is lying on the deck. Prehung exterior doors can also be installed by the rough carpenter. This leaves a complete, closed-in locked-up job.

Heat holes are the rough carpenter's job and should be cut to follow the heat sheet provided. Also, have the carpenter cut the clothes chute hole. The plumber cuts his own stack holes, but a good carpenter crew will often cut them for him. A stack runs from the iron or sanitary line in the basement up through a wall and roof to vent outside. Hence holes must be cut in the wood plates and roof.

Order or put the composition roof on as soon as the rough work on the roof is completed. Do not wait for the plumber or inside rough carpenter work. All this can be finished after the roof is on. Stacks, vents, chimneys, can be done after the roof is on or before. But the trick is, if it rains or snows this work can be completed inside after the roof is on; but if the roof is not on everything stops for the weather.

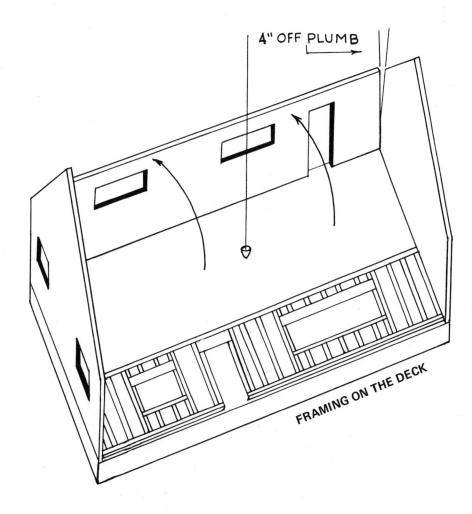

Fireplace dimensions are on the blueprints. If it is a freestanding fireplace, the drawing will show where the prefab chimney goes and the rough carpenter can cut the holes and make sure the chimney pipe clears the joist or trusses. A better deal for a metal preformed unit is to have it on the job and install it while the rough carpenters are there.

22   ROOF

There are several types and weights of roofing to consider. In the country there isn't much wind shelter for a building, so it is good to have some kind of stickdown shingle. This is to prevent the wind from catching the tabs and ripping the shingles up. As the pitch of the roof decreases, it gets flatter, and the chance of leaking increases. A flat roof with no pitch would require a "hot" roof. This is done by applying alternate sheets of felt and hot tar or waterproof compounds. The end result is an on-the-job, manufactured roof. Roofs with less than 4-12 pitch require felt under the shingles. Steeper roofs do not require felt except in areas where snow dams may occur or water may blow up under the shingles. The felt makes a leak harder to find.

Felt on a high-pitch roof, above 4-12, is a detriment in areas where temperature gets above 90 degrees because the roof cannot breath. Without felt, air can circulate between the individual shingles. Also felt frequently wrinkles causing the shingles to buckle.

The roofer will do his job in terms of time the same as other crews, and the more men and mechanization he has the faster he will do it. Union rules often require material to be delivered by truck drivers, then the roof to be installed by roofers. This makes a time lapse of a couple of days as the crew is reluctant to move onto a job unless they have a guarantee that the material has been delivered.

Ideally, the roofers, material, equipment and everything should be sent out on the same truck. This cuts down loss and puts the material in an advantageous position in relation to where it is going to be used. Some delivery truck drivers do not like to drive into a job and up against the house, as there are spots where sewer lines, gas lines, underground electric and water lines are run. These are soft and will not hold up a loaded truck.

Highly mechanized roofers have a powered elevator that climbs up an aluminum ladder carrying half a square of shingles at a time. Here again, team work is important. The truck can be backed against the house and form a base for the ladder. While one man loads the elevator on the truck another unloads on the roof. They can do this for about three squares as too much concentrated weight will warp the roof, or spring some member. To prevent this, bundles can be distributed around the roof approximately where they are to be used. I have seen jobs where five or six hundred pounds of material have been piled in one spot and bent a timber out of line, or broke a porch roof right off.

Actual fastening of the shingle is done by nails or staples. If they are highly mechanized nailers or staplers they will be pneumatically powered by a gasoline air compressor on the truck. People ask if the staples will pull out. The answer is the shingle will tear or rip before the staple will come out.

ROUGH HEAT

After the roof and rough carpenter work is done the next job is rough heat installation. Following a heat plan, the rough carpenter should cut holes and leave openings to install the duct work for a hot air heating system. Beforehand planning is important because ducts have to go from floor to floor between joists and studs. The rough carpenter can best plan on this and adjust his layout to obtain such a result. Colonials and two or three-floor multiples are a little complicated, but the large one-floor ranch allows so much variation in heat layout that often the rough carpenter does not even bother, for the heat man can cut his own holes to his best advantage for comfortable heating. Cold air returns are handled the same way.

Hot water, steam heat, and electric heat require no preplanning by the rough carpenter. The heat man installs his equipment before the insulation and drywall is applied. Other than consulting with the owner as to where radiators or baseboard units are going to be installed, he has no problems.

A clothes chute should be installed at this stage. It involves duct work and can best be done by the heat man.

Transite ducts are heat pipes that go under the cement floors of a building. They are used to heat lower areas, family rooms, basements, utility rooms, where concrete floors are installed.

Heat from these pipes go from a furnace down under the floor, then out registers near windows or outside walls. It is an ideal installation because the heat comes from the floor up. In cold weather areas it is difficult to heat a family room by forcing heating at a ceiling level, trying to make it go down. Thus the transit ducts take the heat where it is needed.

Some communities require inspections for the rough heating installation before drywall or plastering can be applied.

Heat men are good at their trade and today'a furnace units and heating equipment are efficient so that it is almost impossible to get a cold room or an area that will not heat. All that is necessary is to increase the heat pressure to that area.

Modern furnaces are small in size but the heat exchangers are efficient and have no relationship to the size of the enclosure, or box. Thus a 100,000 BTU unit will look the same size as a 350,000 BTU unit. This causes some difficulty in that people feel that the furnace looks too small, without realizing that the exchanger does the work and determines the size in relation to heating capacity.

Central air conditioning should be considered at this point because it requires a different layout for cold air returns if installed as part of the heating system. Most furnaces can be equipped with an attachment in the bonnet that will convert to air conditioning with a minimum of problems.

HEAT DESIGN TEMP. INSIDE 75°F
OUTSIDE -5°F
PROVIDE CANVAS CONNECTORS
USE "EP" TYPE TAKE OFFS.

HEATING SPEC'S

Heat loss 66,757 BTU/hr

Gas furnace JOHNSON HAS 1050 *OR EQUAL*

BTU Input	105,000	BTU/hr
Bonnet Output	84000	BTU/hr

Oil furnace

Bonnet Output		BTU/hr
Firing rate		G.P.H.

Duct work designed by Manual #9 N.W.A.

Bonnet design temp.	160 °F
Bonnet pressure (Total)	.20
Suction pressure (Total)	S. 15 RET. 05
Registers	H&C OR EQUAL

Note: All heat runs to have dampers

Heat loss factors from Manual "J" N.W.A.

Glass	1.35	BTU/hr · Sq. ft.
Walls 1" INSUL.	.12	BTU/hr · Sq. ft.
Ceiling 2" "	.13	BTU/hr · Sq. ft.
Floors BASM'T.	.06	BTU/hr · Sq. ft.
Infiltration windows INCL.		BTU/hr · Lin. ft.
Infiltration doors IN WD5.		BTU/hr · Lin. ft.

ROUGH PLUMBING AND BASEMENT DRAINS

Rough plumbing pipes are installed after the rough heating operation. The plumbers have many opportunities to run their pipes different ways and miss the large heating duct work. However, heating ducts are so bulky that they must be hung in place first because there are only a couple of possible positions.

Plumbing takes two men a day's time on the average one-bath home with electrical power supplied to drill and cut holes for the pipes and stacks. Copper tubing is expensive but faster. This additional cost is more than covered by the saving in plumber's time. Galvanized piping is less expensive, but takes a longer time to install. Copper hasn't been used long enough to really know how good it is. After 30 or 40 years of service, the records will show if it is a better material. As labor costs continue to inflate, galvanized pipe is being forced out of the market. Plastic is showing up in more and more areas. Copper is often stripped from new houses to sell for its scrap value, while plastic pipe is not as attractive for its salvage value. Again, we have no idea of how well plastic will work. Only time will tell, but it is permitted by many codes.

Water tests must be made when necessary as called for in the code. This means that water is pumped into the stacks and sanitary lines to see if they leak while the plumber and inspector are there.

Some areas require basement floor drains; others do not want them. One convenience is a floor drain within 10 feet of the hot water tank, so that it can be drained when necessary if it ruptures accidently, or if the safety valve lets go the resulting water will go into the drain. These drains should be directed into the sewer or sump pump.

Firestopping of plumbing pipes is a must because a fire will travel through the walls and follow draughts. Asbestos cement between the pipe and wood or cement hole makes a good seal. If the space is too large, put wire mesh over the hole, then the cement.

Check with the plumber and list the things he is going to do before he starts, such as: type of fixtures installed, openings for washer outlets, hose openings on outside and inside of house, sump pump and crock, inside and outside drains, how many stacks to put in, and who installs the roof flanges. In many areas, the roofer does the stack flanges for the plumber, by union rules. This means that the roofer must make a special trip to come back and do this small amount of work, which involves only a few minutes. However, it also

61

insures that the roofer is responsible for a leaky roof and cannot blame the plumber.

Sump pump discharge from the house to the drainage ditch is accomplished by using a four-inch plastic pipe. Stretch a string from the sump pump discharge hole to the ditch or discharge outlet location, making sure the string slants down away from the house. Lay the pipe just under the string. In this way, the water will move along the same slant as the string and not freeze. A four-inch pitched plastic pipe will let gravity take the water away from the house and prevent back pressure against the sump pump, which will occur if a smaller pipe is used. Stake out the location of the pipe with wood stakes and a rope or cloth ribbon (a 1-inch bandage will work), so that it can be seen and no one will run over it. Unless a pipe is buried deep, trucks, dozers or cars will break it.

Basement under-floor drains and inside drains are a matter of code in most areas. If the ground is wet after the hole is dug, it can be determined by asking the excavator, people in the area with similar basements, or the plumber if there is a possible water problem with the basement. Do install drains if there is any question about water; even if the code does not require drains. Gravel and perforated plastic pipe are code requirements in some areas.

It is a good idea to check by calling the building department just before putting in the drain, so that you will not be caught in a requirement change by the building code. For example, when you inquired about building it may have been all right to use clay tile drain, now two months later, they require plastic perforated pipe laid in and covered with pea gravel.

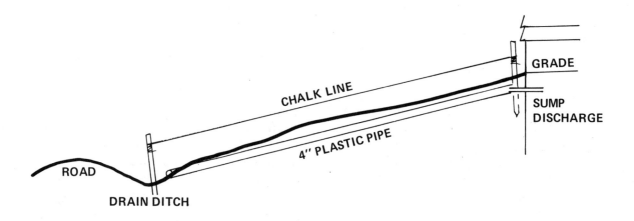

ROUGH WIRING

Rough electric wiring follows rough plumbing because wires can go just about any place, but must go around or over plumbing and heating pipe runs. Wiring must follow codes so that not too much leeway is allowed. Every room has a specified minimum number of base plug outlets, and requires certain size wires. The electrican working in the area will know all of this or will find out before he starts to wire. Many times we have gone ahead without asking and put in the wrong size wire, which had to be taken out and replaced.

Check with authorities in both the subdivision and the municipality to see if underground service is required. More and more communities are requiring all underground wiring. This can be a big problem if left until the last minute, especially in cold weather when frozen ground makes digging difficult.

While checking the above, ask the people in charge of electric service company for a "spot" box location, or where their service wires will enter the building. This location is important because the electric company has a strict code for the wiring to the house, and they understand just where it must be placed. Spot gets its name from a bright orange sticker placed on the outside of the building indicating that this is where the service is to be installed.

The electrician will need a plan or marking on the floor as to just where plugs, switches, and ceiling lights go. The most important item is the swing of a door. Putting a light switch behind a door is a frequent mistake if it is not indicated to the man installing the switch box just which way the door will swing. At this stage of construction there are only rough openings for the doors so he has no way of knowing where the hinges go, except by following instructions.

Plugs in the wall just above the baseboard can vary in position. If possible it helps to spot or locate furniture so that the night light plug next to the bed does not end up on the other end of the wall. A plan we use to accomplish this is to have husband and wife go through the rooms and with large blue chalk marks indicate where switches and plugs will go.

Ceiling lights can be 3 or 4 inches off from the center of a room and no one can tell the difference, but a small variation in such things as lights on each side of a fireplace will show up as a large error because of the nearness of the fireplace, bricks, or trim. Many new homes do not have ceiling lights, but instead use floor and table lamps with base board plugs switched to the room entrances. Thus, when one walks into a room and switches on the lights all the lamps in the room go on. With a three-way switch at the other door it is possible to turn lights on and off without walking back through the house.

In the rough stage, the wiring layout should be checked over carefully if more than minimum code' is wanted. Any additions to consider are best installed at this stage of construction. Yard lights, flood lights, outside plugs for a patio and Christmas lights, inside dishwasher outlets and often outlets for future installations of electrical operated appliances have to be considered.

A word about outside plugs required by code. For maximum safety, they should have waterproof key lock covers to prevent accidents involving children. They will stick garden hoses in them, suck on them, urinate in them when they play around the outside of the house. One case of which I know, involved a chain of children on their hands and knees playing train and the leader or 'engine' stuck his tongue into the outside plug to water the engine.

Inside plugs are available with guards and tops which require turning before exposing the electrical opening.

Electric fixtures have openings into which stripped wire ends are pushed and lock in place to make contacts. Switches and plugs are similarly equipped, no turning screws tight, or making loops, on these newest electrical fittings, thus saving a lot of time for do-it-yourselfers or regular electricians.

LOW VOLTAGE PRE-WIRING

Another rough electrical installation necessary at this point is the low voltage wiring that will be covered up. Intercoms and telephone outlets are the major part of this group. Alarm systems and central vacuum systems are increasingly popular. Some are very simple to install, employing color coded plug-in lines and instructions that anyone can follow, even making future repairs simple for the amateur. Others require a radio-television expert to install and maintain. These are to be avoided at any cost. Television antenna leads require prewiring.

As security becomes a larger item in today's homes, a growing number of electrical monitoring devices are available and should be installed at this point. Censors, horns and electric eye or timer circuits for alert lights are among these. Television cameras are used in apartments to see who is in the lobby waiting to be admitted to an apartment.

Last is the telephone prewiring. This is a system of boxes for telephone jacks placed throughout the house where telephones will be plugged in. Prewiring requires some idea of where furniture will be placed so that a box can be located, for example, next to the bed-stand or on the kitchen counter top, and not back of the refrigerator or stove in the kitchen.

The best type of telephone pre-wiring is a plan using small coils of wire every four feet around the wall of the house. These coils can be energized with a low voltage electric current and the location picked up with a special indicator, outside the finished wall. Thus, the housewife can point out after moving in and arranging furniture, "I would like the telephone outlet about here." Then the installation can be made within two feet of the location by drilling through the wall and pulling the wire out. This sounds quite complicated and expensive but it is not when one considers that inside wall installations serve two rooms. A wall between two bedrooms contains loops every four feet that can be used for either room.

Pre-wiring is not as important on one-level homes as it is on tri-levels, colonials, or multiple units, because on the one-level it is quite simple to just wire from the basement, crawl space or attic, but on other styles it is often impossible to fish wires through floors and fire walls once the building is completed.

Nothing can upset esthetic value as much as a beautiful room with a telephone or door bell wire running across one corner of the ceiling and down the side of the door jamb.

27

INSPECTIONS

Everything is now ready for the drywall or plaster, but first an inspection is necessary or possibly a series of them.

Heat rough inspection

Electrical rough inspection

Sand inspection

Plumbing rough inspection

Drain inspection

Framing rough inspection, structural work.

Care must be taken to check and make sure even if there is not a local community inspection or building department, that the state or some other overlapping government unit does not insist on an inspection.

One area we built in required a state plumbing inspection a week after we started construction. No one knew about it and no one told us about it until it was too late. Then part of the wall had to be removed for the inspectors.

Once these are all okayed, make sure a receipt in writing is obtained from each of the inspectors, before insulation is applied. This could be either a separate receipt for each inspection or the weather sheet signed by each inspector.

Type of Inspection _____	Type of Inspection_____
NOT APPROVED	**APPROVED**
DATE CITY BLDG. DEPT.	DATE CITY BLDG. DEPT.

INSULATION

Insulation must be placed between the studs, with the vapor barrier paper toward the heat or inside of the house. If the paper tabs are put on top of the studs it will cause the drywall nails to pop, dent, or cup the surface of the wall as the paper absorbs and discharges moisture. Also the studs are harder to find, and glue will not hold drywall if applied over paper. Insulation is inexpensive and the difference between the cheapest and best is so little that only the best should be used because the labor cost is the same.

Make sure insulation blankets go behind plumbing pipes to prevent freezing. I have experienced many frozen pipes caused by people packing extra insulation up against a basement or exposed pipe and thus insulating it from the heat.

For best efficiency, stuff all cracks and narrow crevices around windows and doors with insulation pulled off the paper backing.

Ventilate the area between the ceiling and the roof to exhaust excess moisture vapor to the outside before it condenses on the bottom side of the roof.

Ceilings may be done after they are dry walled by laying the insulation, vapor barrier face down, in the joist space, this saves a lot of time.

Wherever there is a high-humidity room, kitchen, bath, laundry, provide a ventilating fan.

Signs of improperly installed insulation or insufficient vents are rusting hardware, furnace parts or nail heads; damp spots on the ceiling, peeling paint, floors bulging, or excessive moisture condensation on the windows, and even walls.

Sometimes people will plug up vent holes with the idea that they are saving heat, but correctly installed insulation will give best results.

Inside walls can be insulated too, especially the inside bathroom wall. This helps to contain all sounds within the room. However, there is a problem: it creates a hush, hush, condition in the home where normal noises that people want to hear cannot be heard. For instance, a nursery can be so soundproofed that the mother cannot hear the baby cry when something is wrong.

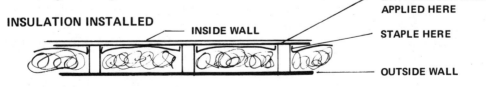

INSULATION INSTALLED   INSIDE WALL   DRYWALL OR PLASTER APPLIED HERE   STAPLE HERE   OUTSIDE WALL

CEMENT WORK

Cement work is priced out by the square foot. However the price varies depending on how much grading and sand fill is required. Most 'flat work', the trade name for cement floors and walks, requires a sand base, and it is often specified as 4-inch thick cement over 4 inches of fill sand. Thus the rate for a simple, flat driveway would be different than a basement floor with over 4 inches of fill sand. It is also worth discussing with the cement man such details as to who furnishes the sand, who cleans out the basement, often full of junk, mud and water, and how the sand is to be put into the basement. It can be dumped right in or pour-loaded through a large upstairs window with a section of the subfloor removed. This way the sand falls right through the joist into the basement. The most costly and difficult method is to hand shovel the sand into the basement through a small basement window. Give all this some thought and direct the truck driver accordingly so that the sand is dumped in a convenient place and will not have to be handled too much. If it is necessary to shovel it through a basement window, dump it right up against the window. Ideally, after the basement drains are put in the sand could be dumped right into the basement before the house floor or deck is put in place. This is important when a large amount of sand is required.

Gravel, large rock sizes, pea gravel, and plastic sheets, are often required under concrete basement floors if a water problem exists. Combinations of these permit water to flow under the concrete to the sewer crock, so that there will never be a water-pressure problem or leaks in the basement. Some building departments insist on having these procedures; others leave it up to the individual. If a spring is located in the basement it should be piped to the sewer or sump crock for relief so that it does not flow under the basement floor.

Keep the concrete ready-mix company phone number handy. Sometimes, it is hard to get delivery on short notice and it is worth while for the cement company to be called in advance for delivery time and date.

Once the job is graded, sand is in place, and sidewalk rails up ready to pour, check to see if an inspection may be required before pouring. Also watch the curb – a 2x4 should be placed against the curb to keep the transit truck from breaking it. In some places there is a $25 fine if you run a truck over a curb without blocking it.

Expansion strips are also required on outside work. The local building inspector will instruct the cement man as to just how he wants them, or if they are not required the cement man can use his judgement. This helps prevent the cement from cracking irregularly, as contraction cracks will occur at the expansion joints.

Once the cement is poured and trowelled it must be protected from children, cars, trucks, and weather.

About the only way it can be protected from children is to stand there and watch it till it sets.

As soon as it has set fairly hard a roll of plastic or tar paper will protect it from rain; straw on top of this will keep it from freezing if the temperature drops at night.

Set up barricades to keep trucks and cars from driving over newly laid cement. Packing crates and sawhorses work well, with string or cloth lines strung around the drives and walks. Depending on weather and what kind of traffic the walk is subjected to, two weeks to a month of keep-off protection is necessary, especially on the corners.

Curb cutting is usually done by the city. Culverts can be installed by the community or the builder can do it subject to inspection and specifications.

A word about freezing weather. Outside cement work is out of the question. Basement floors are done just as outlined. However a working furnace is hung on a shelf from the basement ceiling, or salamanders are placed in the upstairs rooms to keep the heat level up. Chloride can also be placed in the cement causing it to set up quickly and thus not freeze in case the heat is not reliable, or just as an additional precaution. Excessive chloride will cause considerable shrinkage with resulting cracks.

Remove the 2x4 rails about two weeks after the flat work is poured. Some cement men remove them right away, but this is a mistake because the rails offer needed support to the freshly poured edges, which would break off when stepped on if not supported by the 2x4.

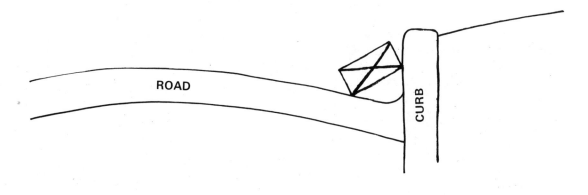

ROAD

CURB

2 X 4 BLOCKING CURB

30

DRYWALL PLASTER

Drywall is used to cover the inside walls. It is faster and less expensive than wet plaster and improving constantly in quality. It comes in various thicknesses — 1/4-inch, 3/8-inch, 1/2-inch, and 5/8-inch. On framing placed 16 inches to 24-inches on centers, the half-inch is sufficient. However, the local code will set the requirements, with 5/8-inch fire-resistant wall board required next to boilers or fire walls next to garages. Some areas will not allow drywall at all, so check the code before starting.

The application of wall board has improved to the point where it is now glued to the framing, eliminating most nailing and possible nail pops. We have been doing this in our area for years. It is very acceptable, and has given results that do not require as much going back to repair nail pops as a result of expansion and contraction of the building materials. The glue webs in between the studs compensate for variations in the framing members. Corner beads then are nailed in place. Taping and cementing the joints follow in order and require quite a bit of skill. Finally, the job is sanded and all slight indentations are spotted or filled in with cement. No further sanding is needed and the job is ready to paint. The sanding and spotting operations are best done after the floor is laid, trim work is in, and just before painting, as carpenters and floor layers sometimes put nicks or nail holes in the walls.

Wet plaster is put on over rock lath nailed and wired to the framing members. Corner beads are nailed on and then a brown or rough coat of wet plaster applied over this backing. Sometimes the community requires an inspection of the lath, wire, and corner beads before the wet plaster is applied. Plaster has various compositions; mason sand was once used on the brown coat. It has now been replaced with an insulation material so that the rough coat has a heat and sound insulation value. After the rough coat has dried, small cracks appear in the corners and around the windows. This indicates the drying has progressed enough; then the white finish coat is added. The quality of the finish coat depends upon the exactness of the base or rough coat, to a great extent.

With trowling and smoothing, the finish coat is completed and let dry. The time it takes depends upon the heat, air circulation and humidity. The finish coat is then ready for paint and is very hard on the surface. In good drying weather, the process takes about two weeks longer than wall board. When the weather is damp and humid it takes a month.

Much depends on how fast the crews follow one another as on the lathing, rough coat and finish coat. Prices are higher, in general, for a wet plastering job than drywall. The plaster contractor has his own lather, material supplier and labor. His price will include all of this. It should be mentioned to him before he starts that he is to clear out the electric outlet boxes after he is finished roughing, as this material hardens like cement and is difficult to remove without damaging the surrounding wall once it hardens. Also, find out about a source of water, as he will need quite a bit. In winter, plastering takes about the same time as summer, but heat must be provided. Either the house furnace or temporary heaters available at rental stores will do this. Once the plaster is dry it will not freeze. Plaster walls absorb sound and noise better than wall board and have a higher fire resistance rating. But because of the cost, time factors and common shortages in good plasterers, drywall seems to be taking over.

Paneling is the third well-known wall-covering system. It is usually reserved for dens, recreation rooms, family rooms, or back of bars. However, we have used it a lot on living rooms, or dining room ends, or on two adjoining bedroom walls, and with pleasing results. Sometimes it is applied over drywall, which has been just nailed in place without tape or cement on the drywall joints. The other way is to lay it right against the framing members. Paneling for this must be 1/4-inch or better in thickness for 16-inch on-center framing. For 24-inch on-center framing, put half inch drywall on first, or use quarter-inch plyscore first, then the paneling. Always place 3/16-inch thick paneling over a backing. Check the local building code. They may want backing even on quarter-inch paneling.

Fastening the sheets to the framing is best done with panel adhesive just as with drywall. It does a beautiful job and leaves a nail free surface. Ring nails in matching color are an alternate method of fastening paneling, or regular four penny finished nails can be used. The heads can be set and the holes filled with panel color-matching putty or wax.

A trim carpenter would quote a price on labor for paneling jobs like the above. The builder or home owner would furnish the material in this case.

Costs of subcontracting the drywall could include all the operations mentioned. So that the agreement would cover half-inch drywall applied, glued, taped, sanded and spotted with corner beads, labor, material, and insurance.

31

FLOOR LAYING

After the drywall or plastering is completed and the house is dry, flooring can be ordered and stacked in the house. Plywood underlay has been more expensive than hardwood flooring at times. Have the lumber company pile the plywood on the floor, then the hardwood on top of it. Plywood seems to attract thieves more than hardwood flooring. The work, noise and time involved for the thief to unpile the hardwood to get at the plywood may help to discourage him.

The plywood subfloor laid in kitchens and baths is to be covered with resilient floor covering or carpeting. Hardwood can be placed in the living room, bedrooms, and dining room. Many home owners prefer plywood underlayment throughout the house, with carpeting, vinyl or resilient tile on top of it. A composition board called underlayment is often used in place of plywood. It is less expensive, has no grain and is lighterweight, but is not guaranteed under linoleum or tile.

In warm weather dry the house out by opening doors and windows until cracks are observed in the edge of plaster where it contacts wood; or with drywall, the taped and cemented seams turn white and hard. Cold or rainy weather requires heat to dry out the house. Once the house is dry, let the flooring age in the rooms for a couple of days, or a week. This helps to adjust the moisture content in the flooring to that of the house.

When nailing the floor down, whether hardwood or plywood, use screw or coated nails with each nail going into a joist or sleeper under the subfloor. This gives the best job. Another system would be to drive staples into the subfloor only. Leave a half inch clearance around the outside of the floor, along the wall. This allows the floor to move, so that if the humidity increases the space is available for expansion. When the floor is laid tight against the wall, any excessive expansion caused by high humidity will cause the floor to buckle.

32

KITCHEN CABINETS AND FINISH CARPENTER

Kitchen cabinets are purchased from a manufacturer or lumber company in units or modules. This simply means that the sizes are standard. You can order them varying width. An example: to fill in a space 3 feet 1-inch wide with cabinets you would order a 2-foot unit, a 1-foot unit, and a 1-inch filler strip to close up the measured distance. Cabinet companies send out illustrated brochures. They are easy to understand and explain their particular systems, which vary with each company. They do everything possible to avoid errors and to simplify ordering. One company makes the cabinets so that they can be turned upside down, thus, it is not necessary to specify a right or left hung door when ordering from them. Another company does not install the door so that it can be hung right or left once the cabinet is in place. Most cabinets are prefinished with stain and a hard laquer spray that approaches a fine furniture surface. Some more expensive kitchen and vanity cabinets are made just like furniture. Try to get the salesman or manufacturer's representative to go over the layout with you before ordering, for it is very easy to make a mistake. Type your order out, as the smallest error in the coded number may cause a different item to be sent out. One order clerk made a written P that looked like a B and an 8 like a 9. Because of these errors, it took two weeks for the cabinets to be delivered, another week before they noticed the order was wrong and two more weeks for the mistake to be corrected, with the house just sitting idle all that time.

One should measure and order the cabinets as soon as the drywall or plaster work is done, but hold the shipping until the carpenter is ready.

As soon as the finish carpenter installs the cabinets, measure them for the counter tops. If they are laminated plastic and straight they can be ordered right from the cabinet shop, using the cabinet measurements as a guide. If the top is U shaped or L shaped, it is a little more complicated and it would be best to have the cabinet man check the job and take measurements. Sometimes the finish carpenter will do this for the cabinet shop. If ceramic tops are used, the tile man will instruct the carpenter just how he wants the tops fixed for his tile application.

When the finish flooring is completed, have the trim lumber and doors delivered, then the trim carpenter can start his work. He should do the cabinet installation as soon as they are delivered so that the counter tops are being made while he is hanging doors and doing the other trim.

Furnish the trim carpenter with finish nails and all the hardware, glue and sandpaper he will need. It is easier to paint if he leaves off the door knobs and catches, but cuts the holes for them. However, if the homeowner is going to paint the house at a later date it would be better for the carpenter to complete the job while he is there. The person painting can remove the hardware for a neater job.

Some trim carpenters can do beautiful cabinet work right on the job. This should not be overlooked, as it frees one from preplanning decisions, in that the cabinets can be made to fit any space and built around the stove and refrigerator, rather than trying to make modules look good with filler strips. Cabinets made on the job usually are made out of 3/4-inch plywood and 3/4-inch finished lumber. This gives the finished product quite a solid, hefty look. It is well built, and is, in my opinion better than some of the low-priced, competitively manufactured cabinets, which tend to be 3/8-inch or 1/2-inch stock. Moldings and decorator hardware, can be applied to job-built cabinets to satisfy every taste.

Stains and hard or soft varnish finishes are applied to complete the job equal to the better manufactured cabinets.

Cabinet installation is very much like building with blocks. They stack up and sit next to each other with everything simplified. Joints are beaded in or have plastic strips, so that any irregularity between modules is covered up or compensated for in a way that doesn't show, even if an amateur glues them together. All this makes for speed and simplicity, but still looks professional when the job is completed.

Also shop built kitchen cabinets installed with glue and screws as soon as possible after delivery have a better chance of not being removed from the kitchen by theft than a pile of cabinet lumber.

In the illustration below the figures refer to the width of the cabinet, thus a V12 cabinet is 12 inches wide, a V-42SD is 42 inches wide.

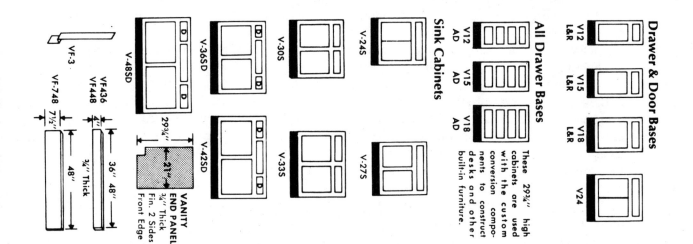

33

TILE WORK

We have found clay tile a popular covering for bath floors and use it on most houses and apartments. It has the advantage of being practically indestructable, repairable, and attractive. A lot of people are used to it, too. Often heard is "My mother and father had ceramic tile. They liked it, I like it, so there is no reason to change." Some of the other coverings are various plastic laminates, and the new brush-on or poured floors and walls. Then there are the resilient floor coverings in sheet or tile form. Vinyl is best. Asphalt tile should not be used in kitchens or baths because it will be damaged by grease or acids.

First let us look at the requirements for the bath room tub or shower enclosure. Decide on how high it should be from the floor. This keeps water from splashing against the wall when taking a shower and every square inch of tile is so much money. Ideally the whole enclosure could be covered, ceiling and all. However, our code height in our area is 6 feet from the finished floor. This works out very well.

The bath floor should, if you have children, be able to resist broken bottles, wet wash rags left on the floor, urine in the corners or around the stool, various hair oil and shaving lotions spilled on the surface, and flooding which will occur if the toilet overflows. Ceramic tile seems to handle all these problems, and more.

However, the other extreme is the placing of indoor-outdoor carpeting right over a plywood or poured concrete floor, in place of tile or other resilient floors.

Some of our bathroom floor codes require a marble threshold raised 1 inch above the finished floor. This is to prevent water from running out of the bath room into the next room in the event of flooding. Another provision is a waterproof cove between the base and the floor. Not a waterproof joint, but a quarter-inch or greater cove which prevents dirt and germs from collecting between the wall and floor. These items may or may not apply to your area. If they are required, you will have to install them. And if they are not required, they do make sense and add to the value of your home if they are put in.

Application of these materials is a professional job. A tile setter takes about a four-year apprenticeship, and as a man works with these materials, he learns the feel of them: at what temperature they will set up, how much humidity the job needs, where to start, how to correct for slight irregularities. I recently learned from a manufacturer distributor that the ink on the quality mark of a plywood underlayment can come through to the finished surface and ruin a poured-type plastic floor. These are things a floor tradesman knows.

Application of these clay tile to the shower or tub enclosure should be over Portland cement plaster applied on wire lath. Water getting back of the tile cannot affect the bond of the tile to the wall. This is the best way. Others are to use adhesive to bond the tile directly to the wall board or to a cement-plastered backing where the cement is applied over rock lath. Without cement over wire lath as a base, the wall doesn't hold up over the years. Many advanced codes require wire lath inspections for tile installation. Preformed plastic bathtubs with 6 foot tub enclosures, soap dish and towel bar in one molded piece now are available.

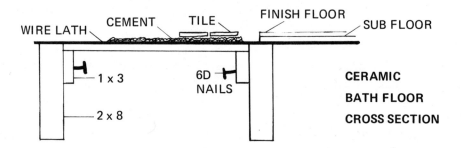

CERAMIC

BATH FLOOR

CROSS SECTION

On clay tile floors, a bed of cement on wire lath is used for a base. If a floor height problem occurs, it can be solved by letting the rough wood floors drop between the floor joist so that the finished floor will not be higher than the rest of the house.

Plastic laminates for wall covering are applied using panel cement and molding, then waterproofing the joints by wiping with a manufacturers specified material.

Resilient floor coverings are applied as the manufacturer recommends and some people can do a fair amateur job on these if everything goes well. Even the new poured floors are coming out in do-it-yourself kits. A lot depends on how fussy one is. A professional does a real professional job and an amateur does the best he can. Some people are gifted at it and take longer but do better than a mechanic, on simpler operations.

Costs of these operations are determined by measuring the job and seeking bids for labor, material, and when can it be done. As an example, specify the pattern and manufacturer of the kitchen floor, how applied, whether the area to be covered includes stair well and landing. Then, be sure to pay between the first and tenth of the month. In case of the bathroom, how is the material installed? Material selected, color, manufacturer? On an item of clay tile there is a significant cost difference between the same color from one manufacturer to another. The least expensive installation is the wall board and the most expensive is ceramic tile applied over Portland cement and wire lath. Cost differences can vary widely because of preferring a certain manufacturer, selecting a special color, or style of tile. Color matching is very difficult. It is hard to pick a tile that will be the same tone as a colored bath tub. After we have selected a color we usually break the tile in two, then keep one half and give the other to the supplier. This way, there can be no mistake as to our choice.

76

34

PAINTING

Painting is a touchy job and requires forethought and preparation. The quality in painting is probably the most important consideration in this operation, and, of course, the better the quality the more the job costs. In my experience, people's judgments vary in terms of quality. I like woodwork in light mahogany finish with a flat varnish that gives it a hand-rubbed look. This is the least expensive wood and finish. Other people feel varnish should have a hard, shiny finish which doesn't appeal to me at all.

Varnished or natural woodwork costs more than painted woodwork. A lot of people tell us just leave the woodwork and paint the rest of the house. From a painter's standpoint, this is more work than just painting everything, walls and woodwork. The reason is that he must keep the woodwork clean as he paints, much of the paint for walls and ceiling is rolled on and the paint splatters around quite a bit. This must be wiped off the woodwork, and the wall paint stopped or cut at the woodwork line, all of which is extra work.

Hardwood floors will be sanded after painting and it is a waste of time to protect them from paint spilled or splattered on them. The same is true of plywood floors to be carpeted or covered with resilient materials.

Standard color paints and flat paint are simplest to apply and the least expensive. The flat covers up many of the irregularities in the drywall. A two-coat job inside and out is sufficient for a new house, for the settling and cracking of the house will usually necessitate repainting and touching up in a couple of years. The first coat is a sealer dyed the same color as the second coat, so the sealer and final coat are all that is necessary to do a good job and cover the walls and woodwork.

The painter will give you a price on the whole job. Make a note of the type of paint and the brand. Check some work he has done, for quality. A professional painter at twice the price of an amateur is cheap. The job will be done faster and right.

35

FINISH HEAT

Installing registers is about the simplest part of finish heating operation, if the furnace has already been installed for cold weather heating. Registers in the floor can go in at any time after the finished floor is in place. Wall vents should be placed on the wall before the carpenter installs the base board which fits up against these registers. Sometimes the furnace is installed on a raised cement pad, which helps in case the basement floods.

At this time preparations for air conditioning must be taken care of if it is going to be done. Control systems, thermostat and all the complete piping and duct work that haven't been done in the rough are installed and tested for operation. It is important at this time actually to fire up the system as the fire pot and duct work have a thin coating of oil or paint on the metal parts. Starting the furnace causes the heat to burn off this oil and paint and it smokes up the house when the furnace is turned on. If the house is not finished this does not damage anything. Just open doors and windows to air out the rooms. But if the house if finished and furniture or carpeting are in place, smoking it up is not good. Some final heat inspections require the furnace to be operating when the inspection is made.

Bathroom registers and kitchen registers in some codes are not to be installed in the floor; they must be installed in the wall. This is a good idea, even if not required by code, because water from a flooded sink, tub or vanity will not run into a heat register if it is on the wall.

At this point we install a humidifier, as mentioned on page 30. A humidifier, when left on during a humid summer, can severly damage a house. Nails and hardware will rust, walls sweat, and mold appears in corners. Of course, a humidifier left on works against an air conditioner, which is a dehumidifier.

Available are power humidifiers installed on the warm air plenum of a forced air furnace. These high-capacity humidifiers are equipped with a drain for automatic disposal of trouble-causing minerals, are identical in construction and features, and differ only in size and evaporative capacity. A motor and fan accomplish the air movement necessry for positive high-capacity evaporation, and uniform, accurately controlled humidity levels are maintained by a humnidistat control supplied as a standard component with each unit.

Programmed thermostats cost about $35.00 extra. A seven-day model wil turn the heat down during the day when the house is empty and turn it up a half-hour before the kids come home from school. Plus, on weekends it will turn the heat on or off.

78

36

FINISH PLUMBING

Finish plumbing fits over the openings of the rough plumbing. The choice of finish fixtures is infinite. The notes on the plumbing contract will indicate the brand name of the plumbing fixtures to be used. However, contracts seldom specify the trim fixtures in great detail. You are entitled simply to a kitchen faucet, a laundry room faucet, an outside faucet, yet quality varies greatly. In case of a faucet, foreign-made products flood the market at times. These are sold with volume discounts and they are cheap in price, although not necessarily in construction. But the low price often tempts the project plumber or large apartment complex plumber to save money by using these off-line faucets.

The main trouble, from an owner's standpoint, is parts. If the faucet needs repair it is almost impossible to get stems, washers, seats or gaskets. Faucets today have removable seats that screw out and can be replaced with a special wrench. Make sure that all fixtures are manufactured in the country in which the house is to be built. Check to see that faucets have removable seats.

Bathtubs in the United States have progressed to smaller and cheaper styles. Our building company gets the feed-back from customers and the follow-up from satisfied customers shows that the majority like the small, shallow tub. In fact, many people order it because of age, arthritis, and other problems. The old style high, deep tub is too hard for them to step into. This is the direct opposite to the idea that if it costs a lot it is best, or if it is huge, big, strong, or massive it must be good. The grab rails around the tub can be installed by the plumber at this time and they should have anchor bolts right into the studs above the tub. These are not required by code in most areas, but are nice to have.

Sinks should be porcelain or stainless steel. An enameled-sheet sink is a waste of money for, in nothing flat, it is chipped and then presents another problem for repair. Stainless steel sinks are growing in popularity as they come down in price.

Present-day hot water tanks have rapid recovery rates and need not be as large as some people think. A 30-gallon rapid-recovery tank does a good job for a family of four, with dish washer and automatic washer. The key to the unit capacity is the recovery rate, not the size of the tank.

Laundry trays are required by law in some areas to be fiberglass. A single laundry tray is sufficient to run an automatic washer by raising a stand pipe system where water is stored around the pipe for the washer. A lot of people do not quite understand this and end up with double tubs which they do not need.

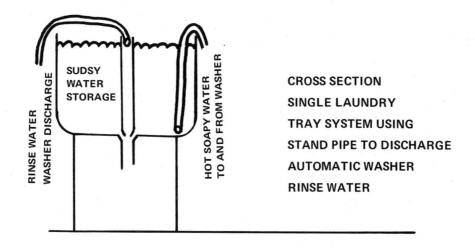

CROSS SECTION
SINGLE LAUNDRY
TRAY SYSTEM USING
STAND PIPE TO DISCHARGE
AUTOMATIC WASHER
RINSE WATER

Garbage disposers are required by some building departments but not permitted in other places. The reasoning for disposers is that their use means less garbage pickup by the city. The reasoning against disposers occurs in areas of septic tanks and bleach fields, and there is the belief that the grinders will load the tanks and fields with ground garbage.

Sump pumps take the surface water out of the basement if there is no sewer. A lot of municipalities where sewers are available insist that only toilet and wash water go into the sewer and that all surface water from the basement must be pumped out onto the ground, into a dry well (a sort of holding hole), or into the street ditch or curb in front of the house. There, sewers just are not capable of taking surface water.

Lift pumps take laundry water from a basement laundry tray to the septic tank or sewer if it is too shallow for the water to run into it by gravity.

These things are good to know beforehand, so that they can be mentioned when talking prices with the plumber. His price may include taking the sewer outside the house, then so much a foot from the house to the septic tank or sewer tap, the place where your sewer meets the main sewer.

37

FINISH ELECTRIC

After finish plumbing, finish electric is next in line. It must follow plumbing because disposers and pumps are to be hooked up, after the plumber, by the electrician. It is very important not to have the electrician come back to do something else because, like all the other tradesmen, he figures trip time on his price — one trip to rough; one trip to finish; and anything left out or not ready involves a call back and more expense. Therefore, everything in finish electrical work should be picked out ahead of time and on the job waiting for him: all fixtures, special switches and location of plugs and switches not previously located in the rough. These are placed after the plumber sets the wash tubs.

Outside electric plugs are housed with waterproof covers so that when closed a hose can be squirted against them and they won't short out. But when opened, the line terminals are available for kids to stick things in or urinate in and therefore can be exceedingly dangerous if placed low enough for them to reach. They can be bought with locks.

Check with the community as to grounding procedure. Some accept clamped to copper water pipes, and other systems involve 12 foot separate ground stakes, because often part of the plumbing water system is plastic. Also aluminum siding should be grounded as it acts as a perfect conductor when a short occurs.

If the electrician can install his switches and plugs before the painter does his work it helps, as the area around the openings needs to be spackeled or painted up by the painter. The plates and fixtures should go on after the painting.

38

LINOLEUM OR FLOOR COVERING

One of the most important ideas to come along recently in flooring is the combined subfloor and finish floor over which the floor covering or carpeting is laid. This results in saving the cost of a finished hardwood floor. This is ideal except that the subfloor used under hardwood is often only three-eighths of an inch thick, while the maximum codes sometimes require 1 1/8-inch for the finish floor thickness. So the plywood subfloor must be 1 inch thick, usually tongue-and-grooved at the joints, and in 4x8-foot sheets. The eighth inch is made up of felt underlay with vinyl or some other type of floor covering, so that the total floor thickness is 1 1/8-inch. The plywood should be fastened down with screw nails driven into the joist underneath. If any of the nails miss the joist, make sure to pull them out, because they will work out through the resilient floor covering after a few years. The joints should be sanded and filled if resilient material is to be applied. With carpeting it is not necessary. Carpeting is being used in kitchens and bathrooms with excellent results.

The actual floor covering depends so much on the underlayment that it is the most important item. Specifications of plywood tell if it is solid. The pressure of a woman's high-heel shoes will go right through a plywood surface having a hollow second core. Thus, the first two layers must be solid, and the up-surface sanded for resilient covering.

Very seldom is a room square, a footing level, a roof straight, or a floor level. Experience has taught that much of it does not make a lot of difference. A four-inch variation from one end of a roof ridge to the other is not visible and not important.

As mentioned before, one of the best talents of a skilled tradesman is to make the job look good visually. However, a noticeably off-square bathroom or kitchen is visual if tile or linoleum with square lines is used; the lines will show up as not parallel to the wall. When one enters the bathroom, where do they look down first? This is the point at which your parallel tile layout should start. As an example, start against the tub to apply the finish floor, and keep it parallel with the tub edge. The other three sides of the room are not as visable as the lines against the tub. Variations can be split between two walls to minimize the off-square problem. In a kitchen, the area for error can be back of the refrigerator and stove, or under the kitchen cabinets, but not a straight wall next to the breakfast table. A non-pattern-finish floor covering, of course, covers all irregular floor covering lines. Building codes often require waterproof underlayment for bathrooms and kitchens. Inexpensive tile or sheet linoleum under carpet will do. Do heat the floor hard-surface covering 24 hours in the room before laying. This keeps it from cracking.

39

SHOE AND HARDWARE

Shoe and hardware are grouped together because it is the end of the carpenter work and involves furnishing the tradesman with all the hardware and shoe for him to install.

Shoe molding is also called quarter round and is a piece of wood that is put down to take up or cover the existing crack between for floor and the base board, or a possible expansion and contraction caused by changes in temperature and moisture content. It should be varnished or stained before being nailed down as the advantage of painting long sticks of wooden shapes before they are fastened in place is obvious. They can be laid side by side and covered with one wide brush stroke, several at a time.

Hardware to be installed at this time includes:

Door handles	Inside shutters
Door locks	Room dividers
Door bumpers	Catches 7 pulls
Door knockers	Clothes chute doors
Door-peep sights	Paneling walls
Stair rail brackets	Closet hooks
Medicine cabinet	Flower pot racks
Shoe racks	Bridging should be checked,
Book shelf plaster strips	along with blocking and
Outside shutters	outside cornice

This is the last call for the carpenter so everything should be checked to make sure he does not miss anything. He is there with all his tools and has them all out for use. For anything on the outside, such as a repair on the cornice; or inside, like adjusting the bathroom door, make sure he is told about it now. Otherwise, it will mean another trip and set-up time to fix some simple thing he missed. The same with material — service him by asking if there is anything he needs a couple of hours before he is finished. There is nothing so troublesome as a missing stair rail bracket and a need for someone going back again to install it.

40

BRICK CLEANING

Brick cleaning is a minor job. However, it requires a professional because there are so many kinds of brick and types of cleaners. Water is necessary and often the lack of it holds the job up until after a well is installed. Many brick companies print in big red letters on their shippers and invoices: "Only detergent-type cleaners to be used on these bricks. Do not use acid as it will bleach the color." Waterlox is a transparent acidproof sealer that can be painted over stone sills or anything that the acid shouldn't touch. Acid will turn a stone sill brown or yellow. Care must be taken not to let the acid run into the basement windows, under doors, or on any newly poured cement. If it does get on any of these flush it immediately with water. The final coat of paint is put on after the brick cleaning, because acid often burns or splotches the prime coat of paint and would do the same if the finish coat were applied before the bricks were cleaned.

Many brick manufacturers instruct the buyer on the exact type of cleaner to use on their brick, and also warn that they will not stand back of the color stability unless their instructions are followed to the letter.

Brick cleaning should be done as soon as possible after they are laid. This makes it easier as the mortar is softer, but it is not an absolute necessity. Jobs can sit a whole year before being cleaned.

PAPER PATENTED BY N.C.R. CO.

SOAP OR DETERGENT TYPE CLEANERS
ARE RECOMMENDED FOR CLEANING BRICK

USE OF ACID IS AT PURCHASERS
SOLE RISK.

ACCOUNTS NOT PAID BY THE FIRST OF THE MONTH FOLLOWING DISCOUNT DATE ARE SUBJECT TO A SERVICE CHARGE OF 1% PER MONTH PLUS LEGAL FEES UNTIL PAID. LOSS OR DAMAGE AFTER UNLOADING IS AT CUSTOMER'S RISK.

THE SELLER HEREIN EXCLUDES AND DOES NOT MAKE ANY WARRANTY FOR MERCHANTABILITY OF THIS PRODUCT, AND THERE ARE NO EXPRESS OR IMPLIED WARRANTIES OF FITNESS.

Rec'd. Above Mat'l. In Good Condition
Rec'd
By

Delivered Above Material In Good Condition
Delivered
By

CUSTOMER'S INVOICE

41

WEATHER STRIPPING AND CAULKING

Weather stripping and caulking are grouped together because the same mechanic sometimes does both. He usually is equipped with a panel truck and carries the tools and material in it.

Outside doors are weatherproofed in most homes to seal them against the penetration of cold, wind, rain, heat and bugs. There are several ways to do this. A bronze spring-type frame is nailed on the two sides and top of the door frame; a hook threshold takes care of the bottom. Another system uses surrounding vinyl beads into which the door is forced, making the fit as air tight as possible. This work should be done last as the threshold and edges can be damaged easily when other construction materials are moved into the house. Another reason to put off the weather stripping to the very last is the amount of moving around a door and frame will do while a house is under construction. After most of the work is done, expansion, contraction and settling tend to have occured and the doors and frames are quite stable and will not require a lot of adjustment.

Windows are weather stripped by the manufacturer. If they are out of adjustment or leaking air because of the weather stripping, the company usually will adjust them. This warranty is limited by time, so if problems develop it is important to make a written request for service before the time warranty expires, usually a year from the date of delivery on the job. However, builders who do a lot of business with them will be able to get some service beyond the year because of repeated orders.

Caulking is done by using a mastic material to make a filler between wood and metal siding or wood and stone or brick. The material flows and adheres well at about seventy degrees. Ocassionally the caulking material will shrink, run or fall right out of the crack. In this case, fill large cracks with paper or wire screen then put a light coat of caulking over the space then come back after it is set to add more caulking.

Caulking goes on best over wood that has been prime painted. This keeps the mastic oil from being absorbed by the wood, whereas if it is put on over raw wood it will dry out and crack quickly. All colors are available and blend in well with redwood or aluminum as an example of two contrasting colors that are carried in stock. Small cracks or irregularities occuring in the workmanship can be covered up faster and better than reworking the job. One example of this is the irregular space between the fireplace brick and the wall molding.

A little caulking will dress up this line to look neat and trim while it is almost impossible to fit the molding to the rough brick surface.

Outside light fixtures mounted on the side of the house should be caulked on the top half, keeping water from entering at the top, and running down in between the siding or brick. Care must be exercised not to "back-caulk" any part of the building. This means do not caulk under something and thus lock rain or condensation water into the building. As an example, a decorative board running horizontally across an outside exposed panel should be caulked only on the top where water will get in between the board and the panel, but not on the bottom, where water, from condensation or driven rain, may run out.

Other fixtures that protrude from the building should be treated the same way. These include, electric wire brackets, meters, mail boxes, house numbers, door bells, and outside electric outlets. Again, caulk the top but not the bottom so that moisture will not be locked in.

A lot of the above depends on the climate, weather conditions and where the building is located geographically. Some areas are so dry that moisture is no problem. Others have bug problems so that every crack must be sealed or screened over.

Just as with weatherstripping, most of the settling, expansion and contraction should be out of the building when caulking is applied. Also, if it is done last, say just before the owner moves in, the exposure for vandalism is cut down. Children take sticks and their fingers to pull the caulking out of the cracks if it sets too long without someone living in the house.

42

CLEAN HOUSE INSIDE

After all the building work is done a cleanup job is necessary. As each trade finishes their job, someone should sweep up, or at least throw the packing cases into the basement or utility room. Tradesmen do not do this because they are costed out at such high hourly wages that they cannot afford to pick up a broom.

The first thing to do in approaching a cleanup job is to call the local fire department and ask about burning the excess building material. Some departments require a permit and allow burning at certain hours. Sometimes someone is required to be standing by with a garden hose or bucket of water. The idea is to burn the rubbish if possible so as not to have the cost of hauling a load of junk to the dump, which could amount to a great expense. Even if the fire watch is not required, it is a good idea.

Choose a place to pile the trash where the smoke will blow away from the house. Start the fire as soon as possible so that it will burn while someone is there to watch it, burning all the large crates and trash first to insure that it will burn out before nightfall. Do not burn if there is a wind, even if the rules permit it. This is too dangerous, and may result in the fire getting out of control.

While the fire is burning, clean the house from the top down. Upper windows first followed by lower window sills, then sweep floors and deposit dirt onto newspaper and throw out on the fire.

Shovel all the rubbish out of the basement and sweep the floor. If necessary, flush the basement with water to get up the fine dust. Do this only in weather when the windows can be opened. In cold weather the moisture from hosing down a basement floor is too much for a new building and will condense on the walls, causing the paint to discolor.

Scrub and clean the kitchen and bathroom.

Install all light bulbs.

Finally put butcher paper or rosin paper in two or three foot wide strips from room to room so that it suggests to inspectors or visitors that they should walk on the paper rather than the floor.

Many communities will not give a final inspection unless the house is completely cleaned up. If it is not cleaned up to their satisfaction they simply write a reject slip and a reinspection plus a fee is required.

43

CLEAN UP AND GRADE OUTSIDE

Cleanup work requires the same procedure on the outside as for the inside, with a rubbish burning permit. Burn everything that will burn. Even big tree stumps will burn to ashes with a hot fire. Green branches will burn, too, if the fire is hot enough. An old tire stuffed under a tree stump will provide enough heat to start the stump or anything else that will not burn easily. Fuel oil and charcoal will do the same thing and cause less smoke and odor. A tree stump, branches, boards, and other trash can be expensive to haul away. Another way to dispose of a big stump or something that will not burn is to bury it in a place where there are no underground utility lines, or where you are sure that a garage or an extension to the present building will not be built. Some communities will not allow junk to be buried, but it is cheap, fast, and does not hurt anything. Grass grows just as good with or without a cement block two feet under it. Another advantage is that if the job needs some fill dirt, burying five yards of junk will save hauling five yards of fill in. In some places, fill dirt is very difficult to obtain, and a wait is involved plus high costs if this is the only way.

Once the outside is cleaned up, dirt can be rough-graded using a dozer or a rubber-tired tractor with a scraper, which will not damage sidewalks or driveways. Cut the dirt down four inches below the finish grade or to your liking. Some prefer two inches. Topsoil or sod will bring the level up to finish grade.

Grading is not difficult but should always have the ground sloping away from the house so that surface water will drain away. One inch drop per foot is a good rule. Keep heavy machines, dozers and tractors and scrapers away from the house. They can do damage to the basement walls and the excessive vibration may cause hairline cracks in the plaster or drywall.

Best results are when the rough grade sets for a while, and goes through some rain storms, but if the soil settles well, final grading and sodding are in order.

Seeding is another way of developing grass. However, in certain areas the cost of sod is close to the cost of labor for seeding. In seeding, a spring rain storm may wash the whole lawn out, while a rainstorm may settle the sod and be good for it. Sometimes, the weather is too dry for seeds to grow, or birds may eat the seeds.

Some building departments require that they set the grade; others leave it up to the builder. Grade is the place at which the ground and the house meet. It can be adjusted in many ways. If it is too high or too low it can be corrected with terraces, retaining walls or air wells. The reason this is mentioned here is that both conditions will occur on a hilly terrain. In Germany, near Stuttgart, the apartment houses are built right into the hills or terrain. In San Francisco, the same thing is true of houses and apartments.

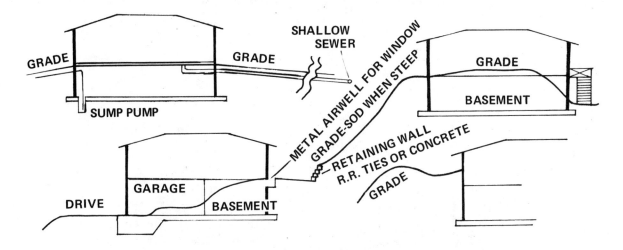

The main thing to watch for is to try to have the grade line above the road, because it seldom floods. If the area is known to be excessively wet, avoid flooding by putting the footer just below the road level and fill in with dirt up to the grade. If it takes too much dirt, build a terrace around the house or along the wall where it is needed.

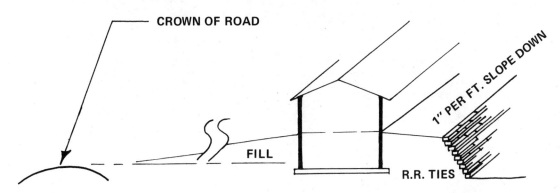

When the grade is below the road, have an air well with gravel on the bottom, or some kind of drain to take water to the sewer, lower road culvert or a sump pump. This also allows windows to be installed below grade.

Retainer walls must be used if dirt or water runoff will go on someone elses property next door or in back. This is code in some places, but also is common sense and common courtesy.

On large sites, grading does not present too much of a problem because of the ample space to work in. On big parcels, like farms, we have even dug huge holes to get enough dirt to fill in around a home to provide the appearance of a gentle slope all around.

All of this involves efforts to control the flow of water around the house. Surface water is the big problem in grading, and if you can control surface water a dry house is usually assured.

Notice use of old railroad ties

44

ECOLOGY

Now comes the Ecology department, requiring in some communities a grading permit before the individual lot owner starts to build. This grading permit and inspection is to assure that there will be no movement of soil off the lot as a result of rain or the vibration caused by machines employed to dig or do final grading.

A flat type of site does not require such a permit except if next to a lake or stream. The idea is to stop the erosion that may occur when the grass, weeds or topsoil growth is destroyed on the lot and rain, wind or whatever is given the opportunity to wash soil into the lake, river or onto the neighbor's property.

Two simple ways to prevent this and pass inspection are first, to dig a small ditch across the lot just back of the water line or across the swale where the water will run. Then, seed this with a rapid growing cover. A second method is to place a barricade of straw bales staked down with steel fence posts, across the lot. This will allow water to run through but hold back soil erosion.

Similarly, as in the case of retainer walls to hold back soil, here we are concerned with holding back soil that water will wash away. This can best be prevented by sowing grass seed even in the crevices and cracks of retaining walls. Gravel stones, and dirt are placed back of the retaining walls and around areas where water will wash through, then seed. Stream beds and sides can be lined with rocks, dirt, seeds and straw to hold everything in place till it starts growing. In time, seeds would find their way into the dirt in these cracks along streambeds. But the newest approach to building before starting is that while one is building, the ecologist wants to make sure the soil is not washed off your property.

Bales of hay or straw half-lapped in a double row as per drawing. Bales must be tied with wire or a non-rotting fiber (nylon). Steel stakes, iron reinforcing rods, steel fence posts or wooden stakes can be used to secure bales in place.

The purpose of the above is to stop erosion and prevent sediment movement off the construction site.

Maintenance on a daily basis in rainy weather is a must; adjustments have to be made for excessive rain or whatever. In areas of heavy concentration of people or traffic, bales are often stolen or moved out of place and, of course, the purpose of this practice is to be ready for rain. Thus, in this case, constant checking is a must before any erosion starts from rain or wash water.

FORMULA FOR SEED TO HOLD GRADE

For 1000 square feet of grass:

one-half pound Tall Fescue
one-half pound Smooth Brome Grass

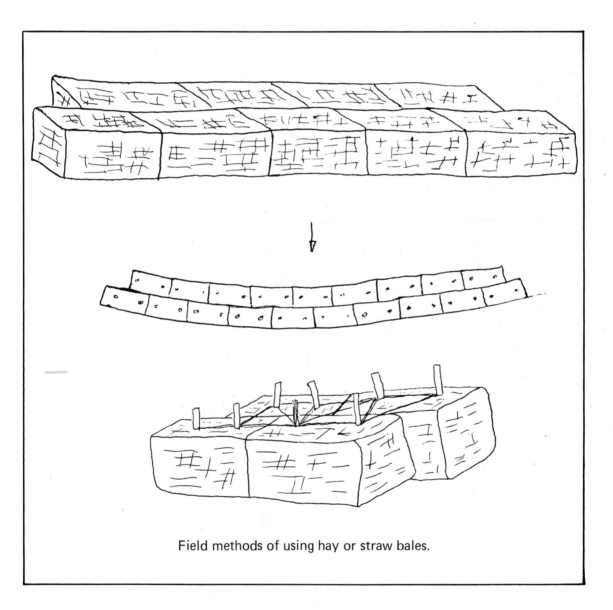

Field methods of using hay or straw bales.

INFORMATION ON THIS SIDE TO BE PROVIDED BY HEALTH DEPARTMENT

ENVIRONMENTAL SURVEY

A. WATER SUPPLY

1. Is onsite well constructed and located according to Act 294, P.A. 1965?_____
2. Bacteriological sample collected:_____Partial Chemical sample collected:_____

3. RESULTS: | Safe | | Unsafe | RESULTS: | Safe | | Unsafe |

4. Is city water available to property?_____
    Recommendations:_____
    _____

B. SEWERAGE SYSTEM

1. Does size of septic tank comply with code requirements?_____
2. Subsurface sewage disposal – tile field – Does it appear adequate?_____
3. Subsurface sewage disposal – other – Describe:_____
    _____
4. Is there evidence of drainage failure?_____
5. What is the likelihood of replacement of drainage system?_____
6. Is space available?_____
7. Are soil conditions in immediate area suitable for onsite sewage disposal?_____
    _____
8. What is the likelihood of a sanitary sewer serving this home?_____
    REMARKS:_____
    _____

C. ENVIRONMENTAL OBSERVATIONS

1. Junk cars on property?_____
2. Are there accumulations of trash?_____
3. Evidence of rodents?_____

Important: Older homes built prior to 1950 could have lead base paints on interior finishes.

REMARKS, OBSERVATIONS OR OTHER HEALTH CONSIDERATIONS:_____

_____
_____
_____
_____
_____
_____
_____

_____   _____
DATE   SANITARIAN,   COUNTY DEPARTMENT OF HEALTH

THE INTERESTED BUYER OF THE HOME HEREIN DESCRIBED SHALL RECEIVE A COPY OF THIS EVALUATION
REPORT.

SOIL EROSION AND SEDIMENTATION CONTROL
ENFORCING AGENCY

SINGLE-FAMILY RESIDENCE
GRADING PERMIT

ISSUED TO: __J.J. Hasenau__ ____ _____
                                            (OWNER)

TO   PERFORM GRADING AND EROSION-CONTROL WORK

AT __Country Club of Ore Lake Sub. Lots 34,35, ½ of 36__
                               (LOCATION)

_____

THE OWNER HAS AGREED TO COMPLY WITH ALL APPLICABLE RULES AND REGULATIONS OF
LIVINGSTON COUNTY AND TO DO NO WORK NOT SPECIFICALLY COVERED BY THIS PERMIT.

DATE __September 24, 1975__   NO. __351 (Three-hundred fifty-one)__

THIS CARD MUST BE CONSPICUOUSLY
POSTED AT FRONT OF WORK SITE

BUILDING IN HOT WEATHER

Most people who have never been in a desert cannot visualize the problems excessive heat can cause. When it gets hot, above 120° Farenheit, it gets dusty. Actually, deserts are green after a rain storm washes the dust off cactus and tumbleweed. Winds up to forty miles per hour can sandblast the paint off of a car and fill a house with dust, an inch thick on window sills. Men can not work in these conditions. Under such odds, it is more profitable to store up basements, foundations or roughed in houses to complete when the weather cools, just as it pays to wait in Cold Weather building when the air is frigid or the roads impassable. To apply the word Desert to all wide-open places is somewhat drastic, but it parallels the 110° Farenheit of Tucson, Arizona or Palm Springs, California where it gets dry, hot and humid. This combination can be a real work stopper, slowing the workmen down, even stopping them.

High humidity is caused by an irrigation of crops in an otherwise non-humid area. Kansas, a relatively low-humidity location last year had the first high-humidity hot summer this year, because of the mass irrigation applied to the corn crops. There are also high-humidity hot natural spots. What this does is slow down construction because men cannot stand the high temperature and high humidity.

To overcome these problems, one could start at 4 or 5 A.M. and end the workday whenever it gets too hot. This is true even in Michigan when some summer days reach 100° where the men are not used to the heat. Farmers use this time-schedule to good advantage through the summer months.

Costs rise in hot weather due to lower productivity, special handling of concrete, lumber warping, and hauling water to control dust. Even the cost of salt pills for the workmen must be considered. Use the opportune time to do as much of the work as possible, so that the later jobs will just fill in the gaps. If you defer the whole job, prices of material and labor will continue to rise.

46

BUILDING IN FREEZING WEATHER

Building when it is frozen outside has an advantage because trucks and equipment can move about. Most people feel automatically that building must stop when winter comes. The biggest problem we encounter is rain and the mud that goes with it, followed by the impossibility of getting from a paved road to the job-site, whether the distance is 400 feet or only one hundred. One can see how a foot thick of frozen ground solves this problem.

In many areas, it really does not get cold enough to stop construction. It would have to be 10° below zero Fahrenheit for two or three days to cause concern.

First, keep in mind when the cold weather starts that there is a certain amount of ground heat that the cold has to overcome before the ground freezes. This heat can be preserved by using straw to cover an excavation, or the area of a proposed digging. Visqueen on top of the straw keeps it from getting wet. Wet straw has no ground cover insulation value. Also, grass and snow will prevent frost from going too deep on ground where you plan to excavate. Be careful not to walk or drive over this ground, as packing down the grass or snow destroys the insulation value. Conversely, if a good frozen access road is needed, to drive over the snow or grass and it will freeze deeper where the tracks are, thus supporting heavy trucks.

In actual digging, a large backhoe shovel can cut through up to a foot or two depth of frozen ground. However, if there are houses nearby, the shock of breaking through the ground could knock the nearest houses down or shake them up badly. If only footings are required, an air hammer will break through the frost. Then work can progress as usual if you follow up by covering the excavation and concrete with straw.

Another main factor in cold weather building is the chill wind index. Bottoms of basements with a foot of straw cover out of the wind in ten below zero weather will be completely frost-free, because they are out of the wind. However, a hill of dirt exposed to the wind with the same temperature will freeze through.

Which brings up the problem of back-filling a basement. It really is not necessary. The house can be built and completed without backfilling until the dirt piles thaw out.

Men can work in zero temperatures when there is no wind. Once the rough work is completed, visqueen can be used to stretch over openings; then temporary heat can be applied to make a fairly warm working condition inside. Outside, visqueen lean-to shelters can be used to shut out the cold. Salamanders with oil or coke heat will keep it warm under these shelters.

As a case history, we once had a thirty unit apartment building which two men completely enclosed by stretching visqueen over all the outside window openings and doors in one eight hour day. A temporary small house furnace was installed and provided enough heat to take off the chill in sub-zero temperatures which occurred only at night. During the day it warmed up a little, and construction proceeded.

Another system is to put in footers or basements and save them for colder weather.

Some subcontractors will work a little cheaper in the winter, some materials are sold for less, which compensates for the added cost of heat and straw. Our particular company seems to feel winter building is an opportunity to beat rising prices, keeps us from being stopped completely by mud and rain, and allows us to have houses ready for Spring buyers.

Visqueen enclosed home 10 degrees below zero

47

BUILDING IN WET WEATHER

No doubt the most important thing to do in wet weather building is to cut drains, so that the water runs to a low point someplace and then can bleed off or be pumped out. This is most important, as is the need for someone to watch the site to see that all drains are free flowing and none get plugged up. We often put pea gravel around the sump pump to keep it from getting jammed with sand or mud.

Slumping of the drainage ditches both on the building site and road drains can be controlled only by constant watching. If the drainage plugs up and floods an area or creates a soft spot, then all one can do is keep at it and wait for a dry spell.

Some sand, when it gets wet, locks together almost like cement. However, a water flow will move it gradually. To handle this, we buy sump pumps with plastic impellers. The sand will cut into brass impellers. Then, if the situation is bad, that is, wet and sandy, we dig the basement a foot deeper than necessary and lay a foot of pea gravel over the bottom of the whole excavation. This, incidentally, has to be dug four feet wider all around because of the banks falling in. Of course, it is important to move right along. Ideally, to dig, put pea gravel and footers in place all in one day.

On clay or mud, the access road presents the same problem in wet weather building since the site must be drained, then the road surface built up with gravel. Check with your excavator. He can tell you just what to use, such as something that will lock together and make a sound surface. Road gravel, or split rock does a nice job. Be careful that you do not get round stones, they are exactly like marbles, or ball bearings, and driving over them offers no support at all. In fact, it makes the situation worse as wheels spin and sink in.

If one has a lot of time at a wet, swampy place, a good road bed can be built up by throwing all kinds of junk into the bed of the road. Then put road gravel on top of this. By junk, we mean broken cement blocks, brick bats, broken concrete or any kind of masonry waste material. We plan the road ahead of time and invite the nearby block yards to dump their yard trash into the low areas. If it is usually swampy and wet, winter time is ideal for this, when the mush is frozen solid.

Dependable weather reports can best be obtained from the local airport. As much as a three day lead can often be predicted with amazing accuracy. Thus you can line up all the subs ahead of time, giving them notice and requesting co-operation on scheduling.

Something that may be helpful to know about is the use of mortar as a drying agent. A basement floor is ready to pour and one corner is wet. Apply twenty or thirty bags of mortar to this corner. The dry mortar will absorb the excess water, and work can continue pouring the cement right over the mortar. All of these techniques for handling extremes of weather are costly, but the work goes on. The house will go up, and waiting in our inflation spiral only adds to the cost. A skilled, experienced builder can tell when to go and when to wait.

Most weather conditions can be mechanically overcome. However, a road ban or road hold condition is a local law, stating that no heavy loads may be carried on a 'no' foundation (as most country class roads are sometimes) from February to May, depending on the local weather condition. Interpreted, this usually means an empty truck or about one yard of material. The local police do not fool around. The fines are so strict that no one can afford to take any chances on violating this law, which is in effect to protect the roads from heavier than automobile traffic. The mushy condition under the surface causes the road to break up even without a heavy load. The road ban will stop construction and there is not one way around it.

Again, an important part to remember in wet weather is to drain surface water into a ditch or sewer. Just hand digging trenches to let water flow from your work area is helpful and will hurry the drying period from several days to one day. If there is water in a basement or a low spot, dig a hole and pump it out into a drain. Once we gassed up a pump and let it run overnight to do the job, then finish the pumping the next day. No one watched the pump overnight, but I would not take a chance like that today. Pumps are too expensive.

VANDALISM

Unfortunately engineers, building estimators and building inspectors do not recognize the seriousness of vandalism.

Again, I have had fifty years of combined experience including my father's, who never heard of deliberate damage. Stealing, yes; vandalism, no. I could write a whole book on case histories, but here are a few of the run of the mill variety. Kids plug up wells and plumbing; they do not realize how much damage a rock or hard brick in a well or toilet hole can cause.

Sometimes the workman you have hired can sabotage the trade job, possibly union or non-union: they did not get the job, the other guy bid lower. One time, a builder's own glazer, mentioned in a newspaper after conviction, needed more work so broke windows on jobs he had just glazed. I have seen electrical damage obviously done by an electrical tradesman, that took just a few minutes, but made it necessary to rewire the whole house.

Then we have the amateur. They just burn the place down or tear it up. One country area had two $80,000 houses burned down in one week. Unfortunately, we had started a $100,000.00 house in the area, and when it was half-built a motorcycle gang kicked the walls in one night. They escaped the local police by travelling across wooded farm fields, cutting fence wires as they went. My customers approached me with the idea that they move in and live in the basement because the weekend fires were making us all nervous. They signed all the necessary papers and as I helped them move in we all breathed easier. Eventually, the authorities caught a drinking husband-wife team who had a weekend hobby of burning down beautiful homes under construction.

A most unbelieveable, fantastic story on experienced vandalism follows. In a built-up subdivision of small houses we had just roughed in a three-bedroom ranch, full-basement house. As the crew and I drove home that night we stopped for coffee on one of the main highways nearby. Off in the distance, the shining new plywood roof was easy to see in contrast to all the darker shingled roofs. It was Friday night. Monday morning I arrived back at the building site. Driving down the street I could not see our house. It was gone. As I got closer I could see parts sticking out of the basement. I discovered that the house had been ripped up and collapsed into the basement, even the forty foot steel I-beam was knocked into the basement. Checking with the neighbors, they said a large gang of drunken men worked a couple of hours doing the damage. However, they were afraid to turn their porch lights on, call the police, or even say anything because the group might retaliate.

100

Our company never has had any labor trouble, so we concluded that a gang had spotted the shining roof and impulsively decided to party it up. The neighbors did sign affidavits to verify to the insurance company that the house was completed as far as rough carpenter work.

All the vandalism in these cases was covered by insurance policies, with the exception of glass replacement. These cases, of course, are the most outstanding and really not bad considering the volume of houses built.

Document vandalism with pictures, police reports and notes. We use a polaroid camera, plus a regular camera, just in in case the regular pictures do not come out well.

Our insurance company does not cover glass, but it will cover window-frame damage. A fire blew out all the double-sealed glass in a window that cost over $1,000. The frame was broken and warped, and I did not take pictures or save parts of the frame. The insurance company refused to pay for the broken frame because I had inadvertently reported only a broken window; they assumed only the glass was broken and refused to pay for the broken frame. Always document the damage.

POLAROID SHOT OF DAMAGE

SAFETY IN BUILDING

In my fifty years of building and being around building, I have been almost fatally hurt on the job twice.

My dad began taking me out on construction sites when I was quite young. Unless you are a builder, I would not allow your own kids to be about the building area. Construction is exciting and draws children like sugar draws flies. My own family is quite used to walking around buildings, but in all the years every one of them has stepped on a nail, probably the most frequent house building accident.

So first things first. Turn boards with nails sticking through so that the point of the nail is down. If a nail should puncture the flesh it may cause an infection, so do see a doctor. Some people become extremely ill, with limbs swelling so they cannot walk. Others have no ill effects at all.

Bottles that workmen leave lying around are dangerous to kids. Collecting a return deposit is one reason. Another is they like to break them against masonry or each other. When they tire of this, vandalism often starts.

Open trenches and shallow two foot deep holes can fill up with water or cave in, making a very dangerous situation for children. Even basements fill up with water. Adults tend to get involved with the construction and not watch the children. In a case, a four year old slid into a three foot water filled hole and drowned.

Electrical circuits for sumps and temporary power are not as safe as permanently installed circuits. And, of course, sump pumps are very, very dangerous. Many people are electrocuted by fooling around with a sump pump in a wet basement. It is necessary to know what you are doing.

A recent news item reported, "A doctor was electrocuted while drilling a hole in his aluminum storm door with an electric drill." It did not elaborate on just what happened other than that it was raining. Imagine the expertise in science education a doctor has, and doing home mechanics he ends up dead.

The percentage of accidents in construction is higher than that of firemen or policeman according to our insurance man.

At times a large timber or pieces of heavy wood are left on the roof around the rafters with the chance of falling off. Hard hats will give a lot of protection in cases like this, or where a tool falls on someone. Most amateurs will put their hammer down. A professional is either swinging it or has it in a holder on his belt.

Nail pounding requires safety glasses, because a spike can bounce up off the edge of a hammer and hit you in the eye. Many glass wearing men have safety glass incorporated into their prescription glasses. Clear glasses are cheap.

A big percentage of building accidents concerns falls. Even experienced carpenters and professionals fall off roofs, step back off scaffolds to admire their work, walk off balconies. It is necessary to pay strict attention to what you are doing, never work when you are tired or have had a drink.

Power tools, including saws, present a special hazard for the inexperienced as well as the experienced.

A number one rule is never work alone. Your wife or children can cover you by just being there. It is a good idea to post the phone numbers of the fire department, police, etc., along with the various building department numbers on the house Permit weather sheet. You might modify this by never working alone on hazardous climbing or roughing jobs that may leave you injured without help. Have someone who can get help if you are hurt. Many trim carpenters work alone, but they have less danger in that the house is closed in and most of the work considered dangerous is done.

One day when I was younger, I was alone up in an attic sawing a 2 x 8 with a handpower saw. The wood was wet and green. Half way through, the saw caught in the already cut kerf as the wet wood closed on it, and the saw took off like a shot, jumping across the floor and across two 2 x 4 studs, ripping them with huge cuts or scars before it stopped. The blade acted as a powerful wheel, digging into anything with its teeth. The momentum kept it going that long. Fortunately, I wasn't back of the saw, but over to one side and it missed me. When one thinks of what that blade did to hard wood, you can imagine what it would do to human flesh. That is why I stress do not work alone.

Remember, these are just a few bad experiences. There are many mechanical procedures done without a scratch. Safety should be an attitude of caution.

CONCLUSION

To summarize, it is the readers choice whether he will undertake to subcontract the house.

Many people successfully subcontract or partially subcontract their homes or additions without realizing that they are taking terrific risks, such as having men work without insurance or proof of insurance or the wrong kind of insurance.

Fortunately for some, nothing happened insurance-wise.

For others not so fortunate, some of the things that turn up are: subcontractors going bankrupt, out of business, or just disappearing. They leave the first-time builder with material and labor claims, and no way to correct the situation. An example of this is that a sub with a shortage of material will take it from your job to finish another job with every intention of replacing it, then go bankrupt, leaving you with a lien for the material, and with no proof of anything, except a debt. There are holdups because of weather, inspections, bank draws, or nonperformance by subcontractors.

Arguments about who is supposed to do something, making a mistake, putting the basement in the wrong place, building the wrong plan, or workmen misunderstanding some part of the instructions, enter in. This book is an attempt to minimize the possiblility of these mistakes and has been developed from experience.

It is apparent that professional builders are faced with the same problems. However they have a built-in cost factor to cover mistakes, a kind of insurance. Each house built has a small amount set aside for problems. Thus the cost of mistakes are taken out of the profit. It has been the author's experience to tear out a basement mistakenly put in the wrong location. But the cost was distributed over thousands of houses. No one can be specifically blamed or charged for some of these happenings. Elaborate systems and check procedures as outlined in the book are developed to prevent errors, but as building steps, materials, and procedures get more complicated the chances of costly problems increase.

Actually you are dealing mostly with people and, as the saying goes, "People are people." It is easy to make a mistake without even trying.

So, add a little, say five percent, to your top-estimated total cost for problems, and good luck!

INSTRUCTIONS: **SCALED PLOT PLAN FOR PERMIT NO.**

Sec. 3.2, Article III "Sanitary Code",   County Department of Health requires scaled plot plan providing the following information.

(1) Lot Size _____   (2) North _____   (3) Fronting Roads _____   (4) Grade Changes _____

(5) Any Easements _____   (6) Building Location, Size _____   (7) Water Well Location _____

(8) Septic Tank, Drain Field _____   (9) Water and Septic Tank Systems — Adjoining Properties _____

(10) Fronting Lake - Stream _____   (11) Driveway _____   (12) Utility Lines _____

Please Note: Subject property will be identified with boundary markers.

Recommended Scale ¼" = 10' — If otherwise, please indicate.

250
240
230
220
210
200
190
180
170
160
150
140
130
120
110
100
90
80
70
60
50
40
30
20
10

10 20 30 40 50 60 70 80 90 100 110 120 130 140 150 160 170 180 190 200 210 220 230 240 250 260 270 280 290 300 310

Name of road (not to scale)

Lot No. _____ Subdivision _____ Signed _____

City, Village, Township _____ Date _____

188 Pgs. Soft Cover
Illustrated
8½" x 11". Indexed.
$12.95

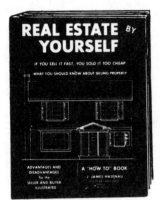

124 Pgs. Soft Cover
Illustrated
8½" x 11". Indexed **$12.95**

106 Pgs. Soft Cover
Illustrated
8½" x 11" Indexed
$12.95

BUILDING CONSULTANT   J. James Hasenau

THE OWNER'S GUIDE TO UNDERSTANDING CONSTRUCTION OF HIS HOME.
WHAT YOU SHOULD KNOW ABOUT BUILDING SPECIFICATIONS.
ADVANTAGES AND DISADVANTAGES FOR THE HOME BUYER.

An Encyclopedia of building language that every prospective home owner must know.

Things everyone should know who plans to purchase a home or have a home built, whether you use a builder, sub-contract the work, or do it yourself with your own hands.

The author, a second generation builder, who has built hundreds of homes, tells the home buyer about the bad builder and the good builder, and all about construction. He describes the entire building process including the selection of a builder, and step by step what he should do. He tells the buyer things he must know and do to get a first class home built.

The book talks of the pitfalls which cause the most heartaches and disputes during the building process and what to do about your mistakes or the builders errors. It tells how to avoid serious trouble and extra expense. It will give you understanding of the fundamentals that every home buyer should know regarding construction of his home.

You will earn money saved, not only during the construction period, but also for years afterward because your house will have good planning, good materials and good workmanship. You can avoid costly errors.

If your house is already built you should know how the house you plan to buy has been built, what kind of material has been used, what to watch for.

With this book you are your own builder consultant.

The contents include:

The bad builder	Plans and specifications	Decorating
The good builder	Fireplaces	Windows
Mistakes	Building Site	Clean specifications
Changes	Floor framing	Partition framing
Choosing a builder	Lath-Plaster Drywall	Exterior walls
Lawyer	Ceiling framing	Arbitration

REAL ESTATE BY YOURSELF   J. James Hasenau

IF YOU SELL IT FAST, YOU SOLD IT TOO CHEAP.
WHAT YOU SHOULD KNOW ABOUT SELLING PROPERTY.
ADVANTAGES AND DISADVANTAGES FOR THE SELLER AND BUYER.

Some of the how it is, front-line action in the Real Estate business is told in real life cases that the author has experienced. Dirty tricks and emotional involvements in sales techniques are quoted. The book does not have cut-and-dried simple sale by advertising or selling; it is much more complicated today. Learn what is is all about. And do not sell too cheap.

The contents include:

Dirty deals	Deposit	Basic selling procedures
Shills	Sunny side up	Qualifying the buyer
Appraisal	Impulse buying	Mortgages
Down market	Dealing with defects	Land contracts
Fixing up	Basic bargaining	Closing
Advertising	Rent or option	Understanding the legal

RENT   J. James Hasenau

A GUIDE FOR HOUSE AND APARTMENT OWNERS

THIS is one of the best money-saving ideas ever to be published on RENT, and it will answer the many questions that arise in a tenant or landlord's mind as to how to handle a certain situation facing the renter, or the owner landlord.

Although many incidents may have a familiar ring, the answers are clearly set out in a practical way. The background of both the author and his wife cover this field thoroughly. They have done this through rentals of every type involving a tenant relationship, with apartmenbts, duplexes, and single homes.

Therefore, the material is fresh and up to the minute.

Among the many topics sthoroughly explored are: setting the rental amount; security deposits; leases; screening; collecting rents; returning security deposits (time lapse required); including applications on how to keep records; how to handle rent strikes and avert them; insurance extras; advertising; choice of tenants; security for vacated property; security for occupied property; investing in rental property; trouble calls; service people; tools for repairs; paint; carpeting sources; garbage disposals; exterminators; preventive maintenance; and keys.

This is a reliable guide and is recommended to all who are interested in being a tenant, or who may wish to know how the landlord handles certain situations from the tenant's standpoint. It is a book you will need, even if you are a landlord for only one home, store, or building.

The contents include:

Setting rental price	Trouble calls	Check sheet	Security for vacant property
Lease	Simple repairs	Screening	Security for occupied property
Screening	Service people	Choice of tenants	Investing in rental property
Applications	Taking over after vacancy	Collecting rent	Professional management
Security deposit	Insurance	Records	Tool box

--- ✂ ---